A Hive of Bees

JOHN CROMPTON

ILLUSTRATIONS BY A. E. BESTALL

NLB

Nick Lyons Books

Printed in the United States of America

10 9 8 7 6 5 4 3 2 1

Published by arrangement with Doubleday & Co., Inc.

Library of Congress Cataloging-in-Publication Data

Crompton, John. 1893–
A hive of bees / John Crompton : illustrations by A.E. Bestall.
        p.   cm.
Reprint. Originally published: Garden City, N.Y. : Doubleday, 1958
ISBN 0-941130-50-9 : $8.95
1. Honeybee.   2. Bee culture.   I. Title.
SF523.C857   1987
595.79′9—dc19   87-17623
                    CIP

CONTENTS

I am going to quote John Crompton sooner or later, so it might as well be sooner. Certain pleasures you don't want to delay. "At the present time," he says in his own introduction to *A Hive of Bees*, "it is reckoned to be one of the cardinal sins to attribute human qualities to non-human animals, but for the life of me I could not help doing this with my bees…if you wish to call me anthropomorphic you may do so. A mere word, even one as long as that, never hurt anyone."

Reading *A Hive of Bees*, you will surely call Crompton anthropomorphic. He cannot talk about bees (or spiders or snakes) without using human analogies. And it may occur to you—having discovered the fun of Crompton on the hive—that one reason anthropomorphism gets condemned is because, unlike Crompton, most writers who indulge in it are mediocre anthropists, no matter how good their science. They forget something Crompton knows very well: anthropomorphism is a two-way street.

Crompton understands the limits of what Gilbert White, the great eighteenth-century British naturalist, calls "analogous reasoning," which is the fallacy behind anthropomorphism. He does not presume to make great discoveries by likening bees to man, nor does he claim to be a scientist or even a practical advisor for beekeepers. He simply tells stories about bees which

we have heard before with men, women, and nations in the
leading roles. For the uncertain reader (certainly not for the bee),
Crompton lends *Apis mellifera* some human ethics, a little myth,
and a little history. The results are enchanting.

Consider perhaps the finest example of Crompton's anthropomorphism in *A Hive of Bees*. He opens a hive for us, and we
discover that its internal workings are incomprehensible, mysterious. All we notice is a buzzing blob of bees doing delicate things
with tiny mouths and glands. Then he shows us the outside of
the hive.

> This is our stage. Here we see the play: those coming
> and those going, those staggering in with sacks of
> pollen—the bread of the community—those laden with
> honey, those bearing water, those carrying, with ex-
> treme care, cargoes of the dangerous, sticky propolis—
> cement for the masons with which they reinforce the
> flimsy manmade hive and bind it down as if welded. We
> see the drones, the bucks of the community, swaggering
> forth, dandified, handsome, brainless like their proto-
> types, the Corinthians of old, and like them, contemp-
> tuous of any honest work. We see the children in their
> new gray velvet suits brought out by their nurses,
> sitting in the sun and waving their front legs enthusias-
> tically to the swift, unheeding traffic that passes over
> them. We see the guards, conspicuous as if arrayed with
> swords and helmets, at their posts near the entrance,
> challenging every comer. Yes, it is much more human.

In color and movement, this resembles fine nineteenth-century
narrative history, say, Prescott on Peru or Parkman on Quebec,
for Crompton here depicts the daily life of a great ancient city of
some fifty thousand inhabitants, which is exactly what a hive is.

Of course, the bee is a traditional subject of anthropomorphism, easily adduced, easily exampled. She (and bees are by

number overwhelmingly female) is the source of sweetness and light, the emblem of industry. She has that barb to remind us of her aloofness from man. To our stock aphoristic image of the bee Crompton adds a surprising dimension of moral psychology. Here, for instance, is what he makes of the "busy bee": "Idleness is probably more demoralizing for the bee than for any other creatures." "The bees' religion," he writes, "is posterity." He says of swarming, that complex moment in which the oldest members of the hive depart with the old queen, leaving the hive in the possession of the young, "sometimes it seems as if the swarm takes with it the brains of the community.... Possibly they think this only right. They bequeath to those they leave their city, their wealth, their children—everything. They ought not to be grudged the wisest members—they will *need* them."

This is not *mere* anthropomorphism. It is good amateur bee science tempered by the pathos of human history. And it only goes so far. Behind Crompton's dramatic rendering of the hive's life lies the paradoxical sense that human analogy works precisely because bees, and all non-human forms of nature, are so alien, so remote from our experience. "You see," he says, "I wanted to know *all* about bees. My efforts only brought me in the end to that impenetrable curtain that shuts off from the gaze of man the life of the bee. He may see so far, but no farther."

I have been talking as if *A Hive of Bees* turned on the question of anthropomorphism. It doesn't. It turns on the good humor, the quiet irony, and the solid information Crompton brings to a wholly fascinating subject. Insofar as the wonder of beekeeping can be explained, the author of *A Hive of Bees* does so. The image one has of Crompton after reading this book is that of a man standing hypnotized (cigarette in hand) before a white box squatting in the grass.

That is an accurate picture of the amateur beekeeper. I know. When I was a boy in Iowa my father kept bees in an asparagus patch in the garden. For several nights before the first stock of bees arrived, my brothers and sister and I assembled hive parts (brooders and supers and combs) in the basement. These we obtained from a bee supplies dealer in Sioux City. Then one day the bees arrived and were carried to the garage where we fed them and calmed them. They came in a wire frame box with a slender detached suite of rooms for the queen and her aides. The entire contraption was just big enough to hold two laying hens. It was like ordering a dairy herd and bull through the mail. As the season wore on, we learned to uncap honeycombs with a hot knife and centrifuge them in a large spinning tub near the washer and dryer. If you turned a tap near the bottom of the centrifuge, you could fill bottle after clear bottle with pure Wright County honey. For us kids, it was, as Crompton says, "such money for jam." For my father, who wielded the smoker and who wore the white bee suit, it was a different matter, but he too was absorbed in the hive.

Crompton has got it all exactly right, but with a characteristically English touch. Beneath the flights of foraging bees, you can discern in his book the quiet landscape of the West Country, the gardens and copses and the hedgerows that look like long lines of low beasts moving slowly across country. When a bee settles onto a flower, you can hear Crompton's voice saying,

> The two are wedded. It was a marriage consummated in a dim and distant past, but they have been faithful ever since. Neither could survive without the other. It is no Platonic union: the virgin stigma of the blossoms knows nothing of the distant anthers. The bee is *her* husband. He comes to her lusty and pollen coated, and is received with open arms. It is a happy, fragrant marriage.

This is enough to make you wish that Crompton had written a book called *The Human.*

—Verlyn Klinkenborg

INTRODUCTION

Rarely has the acquisition of a hive of bees had such a drastic effect as in my case. It made me neglect my work, drop my studies, leave the garden to its weedy fate, arrive late for meals and, worst of all, it resulted, later, in this book—for surely there are enough books on bees already. There are so many, in fact, that most writers in this crowded field start with an apology, then make excuses and give reasons why their books fulfill some want. I am quite ready to apologize, but I make no excuses for the simple reason that I cannot think of any. Still, unlike some books, a book on bees does no actual harm and no one is forced to read it.

Practically all my life I have been interested in insects and many other forms of life, but never, until one memorable day, in bees. The reason for this change of front is told in the first chapter so I will not go into it here. I should be ashamed to confess the amount of time I spent watching bees during the seven years I kept them. Some people absent themselves from work for the excitement of watching football matches: I got just as much excitement watching bees—more, in fact, for I do not get very excited watching football matches.

Then the war came. I joined the R.A.F. and went on courses

and the usual intensive training, emerging as a flying control officer ready to be posted to his first station. There was an airdrome near my home called St. Eval. If I could be posted there I would be able to go over and attend to my bees in my spare time. So I applied for this posting. At the same time my chief friend on the course, a Scot, applied for a posting to an airdrome in Scotland. It was the wrong thing to have done: I was posted to Scotland and he to St. Eval. It was sometimes possible, however, to arrange an exchange. We tried this and the request was granted, then, two days later, canceled. So my friend went to St. Eval, and on his first night on duty was killed by a German bomb, and I went to Scotland.

I did not stay there long. Perhaps the authorities thought this was not far enough away from my home. I was posted to Iceland. And here it soon became obvious I was likely to stay for some time. Flying control officers were in short supply and fighter and bomber planes from America were being flown over to Britain in large numbers, via Newfoundland, Greenland, and Iceland. Three of us, exchanging twenty-four-hour watches, had to cope with all these, land them, and send them off again. I hardly had time to be worried about anything except airplanes, but I did begin to be worried about my bees. The war did not seem to be ending, indeed it seemed hardly to have begun. There was only one thing to do. I wrote to a beekeeper friend near my place asking him to have my bees sold—the whole lot, hives, equipment, everything.

I got a letter from him later saying he had done so and had got a very fair price. I did not care about the price. My bees had gone. I felt like a father who has lost his children—the whole 500,000 of them.

Shortly before the end of the war I was invalided out of the Air Force. I did not take up beekeeping again. I had no hives or equipment (and the price of these had risen enormously since I had sold mine), my special strains of bees were gone (like everyone else I always thought my strains better than any other strains), my disability made it impossible for me to lift heavy weights, and I had had to leave my home for a place where there is little fodder for bees. Where I am now is a rich district in many ways. Hops, wheat, and beans do well, but crops like these, which spell wealth for farmers, spell poverty for bees. Not very far away to the north are extensive cherry and apple orchards—a veritable promised land for bees, but just too far away for any that live here.

No doubt, too, the first fine frenzy had departed. Indeed, it could not possibly have lasted at the tempo with which it started. And I had plenty of other insect interests. But I still keep up with bees, read most of what is written about them, go to meetings, swap yarns with beekeepers, and frequently wish I had my bees back again.

This book is a mixture of experiences, impressions, and speculations as well as bare facts about bees. It is not a textbook; there are no tips for commercial beekeepers and no biological data for students. It is, I think, up to date in the essential details at the time of writing, but even with an animal that has been so long studied as the bee new interpretations of its behavior are always coming forward.

So are new gadgets. Only the other day I heard of a new device that was being tested. It is a sound device on, I think— I am very ignorant about these things—the radar principle. It can penetrate an unopened hive and send back the sounds the bees are making inside, for it is thought now that bees have a sort of language. For instance, a certain note will indicate that a colony has decided to swarm. There will be other notes of course, sounds of anger, sounds of mere contentment, and the rest, but the swarming note can be sorted out, and if it is there the beekeeper can take his anti-swarming steps. Last week I heard an expert explaining this and other new things to an old beekeeper. "The time is coming," said this expert, "when mistakes will be eliminated and the whole business made a certainty. What a boon that will be for you beekeepers!" "Ye-s," said the old beekeeper doubtfully, "but all the fun will have gone out of it."

There is a lot in that. Women keep our interest because we never know what they are going to do next, and bees are the same. But I am not worrying. I do not believe that all the scientific gadgets ever invented, now or in the future, will make bees change their unpredictable ways.

At the present time it is reckoned to be one of the cardinal sins to attribute human qualities to non-human animals, but for the life of me I could not help doing this with my bees. We do not know, of course, what the mental processes of other animals are but, with insects at any rate, we can be quite sure that they are not like our own. Yet in this book I have sometimes spoken of overseers, councils, committees, and so on and so forth. Do not deduce from this that such exist in a hive, or at any rate that if they do they conduct their affairs on human lines. (They could not afford to, for one thing!) I bring them in only as analogies to clarify and simplify the picture of bee life that I try to give. These imageries helped *me* to understand and keep up interest in the bees I watched, so they may help others. In any case, we have no clue to the nature of the organization of a hive and, lacking that, must fall back on what we *do* understand.

But if you wish to call me anthropomorphic you may do so. A mere word, even one as long as that, never hurt anyone.

CHAPTER ONE

THE CONVERSION OF A BEE HATER

I used to live not very far from an agricultural college called Seale-Hayne, and passing by one day, I thought I would like to have a look round. It was vacation time, so I would interfere with no one. I saw the caretaker and he gave me permission to go where I liked. I made a sort of tour of the place, and on my return stood chatting with him. Then, to my surprise, people began coming in through the imposing outside gates—elderly women and elderly men and a few younger ones. Talking earnestly together in little groups, they disappeared into the building.

The caretaker never gave them a glance, and I asked him who they were. "That lot?" he said. "Oh, they've got bees in their bonnets."

"You mean they're mental?" I asked, shocked.

"Well, not exactly," he said. "There's a bee course here twice a year—lasts a week."

"You mean they come to learn about bees?"

"Yes."

"Oh . . ." I said. "Well, I've already learned all I want to learn about bees, and a lot more than I want."

He said he had, too, and with that we changed the subject.

Little did I think that in but a short time I also would come walking through those gates, bound for some classroom and talking earnestly with others—about *bees!*

I often wonder where the fascination of the bee—to those it does fascinate—lies. No creature has had more literature devoted to it: a continuous outpour, from Aristotle and Virgil down to the present day. It is a remarkable insect in the way it arranges its affairs, we all know that. But so are ants. It is a familiar of summer days: our flowers would lack something vital without the song of the bee. That tends to make it popular, I dare say. By pollination it gives us our fruit and our clover fields. And it makes honey. Honey is good, and there is something intriguing, almost impious, in eating the distilled essence of the living flower. These things count, but do not *quite* explain.

For whatever reason people *do* like the bee, I was not one of their number. To be candid, I hated bees. The hatred arose in Africa. Bees there were wild bees, in every sense of the word. The buildings we mostly knew (when they were not pole and mud huts) had galvanized iron roofs with match-boarding underneath. Into the hollow between, wild swarms often came—a highly unsuitable residence in that hot climate; more unsuitable for them even than it was for us below, but that was their affair. The trouble was that, having established themselves, they resented to the utmost our presence in our own home. African bees have a much hotter sting than other bees (or so it seemed), and are very much readier to use it. I remember as many as six at a time sizzling down like little red-hot cinders into the roots of my hair. They particularly used to dislike a sweating horse or mule, and when a swarm had taken up residence in our camp it was no uncommon thing on arriving home hot and tired to have one's mount kick up its heels and bolt just when one was halfway through the act of dismounting. This provided amusement for all except the horse and rider.

Naturally the bees had to be dislodged from their quarters, which made for more unpleasantness. So a feud grew up between bees and ourselves. We disliked them, and they were far from loving us. I must admit they had a certain amount of justice on their side. In the veld whenever a bees' nest was found (and the little honey bird would generally show us quite a number during the course of an average patrol), our sole idea was to ravish it. Trekking in the heat, one craved for sugar. The natives did the ravishing. It was rough and ready, but thorough. The bees were in no condition to lodge protests afterwards. And such were my feelings towards them that no qualms of conscience troubled me. In fact, I felt almost as pleased at having got some of my own back as I did in feeding on the honey.

Eventually after twenty years in Africa and China, I had to go to England for health reasons. Like many another returning wanderer, I wanted a small place in the country with a bit of ground. I got one. (I got stung again, too, but not by a bee this time.) Having got my place, like all the rest, gardening began to appeal to me. I did most things that gardeners do, and finally became absorbed in fruit trees to the exclusion of the rest. The fact that a couple of apple trees in my garden never had any fruit started me off. I inquired into the matter and did things to them. I studied the subject. I got pear trees and more apple trees. Then I branched out into plum, cherry, and peach trees. The birds took what cherries there were, so I made a cage. The little bushes went in easily enough, though in a few years they were trying to push the roof off.

I had succeeded in making my pear and apple trees fruitful (or they had become fruitful in spite of me—I don't know which), but I was not so lucky with the others. There was a mass of blossoms every spring, but very little fruit. I forget how much the cage cost me. At least twenty pounds for the materials alone. So, ignoring the cost of time and labor in erecting it, I had paid twenty pounds to preserve intact two, or possibly three, cherries per annum. I did not need to study interest tables long to find out that my venture was not financially sound. I thought the matter over. The trees blossomed well. The flowers were not killed by frost—at any rate, they bore just as little when there were no spring frosts as when there were. They were of different varieties, so lack of cross-pollination was not responsible.

Or was it? Did not bees, and bees alone, effect pollination? My apple blossoms in May and June had been alive with bees, but now I came to think about it I could recall no bees on the plum or cherry blossoms. Yet someone in a gardening magazine had said that bees went crazy over cherry blossoms. Why had not they gone crazy over mine? I made inquiries. I found the nearest "apiary" (a word I had not struck before but which apparently denoted two or three hives gathered together in one place) was a considerable distance away. This seemed hard. The supply of fools is proverbially unlimited; in fact, they are supposed to be born at minute intervals. Yet no one fool enough to keep bees was located closer than two miles from my garden! For I was on the right track. Apples know the fickleness of spring and take no risk. They blossom late. And when they blossom, the local bee is in a position to attend to them. Hives have had time to build up after the winter, and days are warm enough for distant flights. Cherry and plum trees come from Asia or some similar place and blossom when their remote ancestors blossomed, at a time before our bees have got into their stride. This is all right provided they are fairly close to hives. But if it means a long journey, bees cannot reach them in chilly weather.

It almost seemed—and the thought was most disconcerting—that I would have to get some bees and put them on the premises. It shows how keen I was on my fruit trees even to have contemplated the idea. Normally I should have counted myself lucky in having the nearest bee so far way. But I was determined to get cherries. Also, that twenty pounds rankled. So I made the decision, with reservations. I would get bees, but that would be the beginning and the end of it. I would put them in one of those straw things, near the cherry trees, and leave them alone. They could have the run of my garden and they could keep any honey they made. All I asked of them was to pollinate my fruit trees. It was to be a sort of gentleman's agreement. What I hoped was that they would realize this and not (as their African relatives would have done) ask me pointedly what I was doing in *their* garden.

How one acquired bees I had no idea. Quite by chance, a week later, I was introduced to a man and learned he was a bee enthusiast. I cultivated him and found him quite normal. He did not even mention bees. Finally I approached him and laid my cards on the table. We were at cross purposes from the start, though neither realized it. The fact that I wanted bees only to pollinate fruit trees never sank in. To him it was that I had seen the light and wanted bees. He said I could not keep them in a straw skep, and I took this literally. He had a hive, an old hive but serviceable,

and this he gave me. He also gave me a list of things I would
have to get to put inside. These were numerous, unintelligible
and, as I found afterwards, expensive. When I had got them he
promised to come over and fit them up inside the hive.

He did so, and it all seemed perfectly crazy when he had
finished: the sort of interior Heath Robinson would have made
for *his* bees if he kept them. But my friend was satisfied, said that
all was now ready, and that what remained was to get the bees.
The hive looked so full of hanging frames and other stuff that
I could not see where the bees were to go, unless they lived outside, but I kept my doubts to myself. My friend said it would be
best to get a swarm. He could not let me have one of his own
because he never had any. (All beekeepers think this. They have
infallible, non-swarming systems, so that only the bees of other
people swarm.) But he said he knew a man who had a swarm
in a box, and we arranged to go there straightaway, calling at his
apiary to get a few necessities on the way.

The first thing we saw at my friend's apiary was two swarms
hanging from bushes near the hives. I said it saved a lot of trouble;
I could buy one of these and the man could keep his swarm in
his box. He stared at the pendant masses and bit his nails. I saw
that parting with either of them was going to be a wrench. He
hummed and hawed, and finally said he could not spare them. So
we went to the man with the box. The man was there and so was
the box, but the bees had gone, and after seeing the box I did
not blame them. The man said his neighbor had had a swarm
yesterday and might be willing to part with it.

The neighbor had his swarm also in a box, and was willing to
part with it. Moreover, he insisted on giving it to me for nothing.
I argued, but he was firm. I believe that, like most country people,
he thought it unlucky to accept payment in cash for bees. The

box lay upon a slab of wood and had a small hole bored in one side, into which an occasional bee entered in a halfhearted sort of way. My friend picked it up and turned it over and looked inside. As soon as I saw what he was about I went quickly over to examine a small laurel tree growing at the far end of the garden. However, the bees did not seem to mind. A few came out and hovered around his face as much as to say: "I say! Dash it all!" But they soon went back through the hole when the box was up-righted. We waited a little and then clapped perforated zinc over the hole and secured it with drawing pins. After which we tied the box firmly down to the slab of wood and carried it to the car.

The bees were now waking up to the fact that something was amiss, and that they were imprisoned. Imprisonment sends bees crazy. Imprisonment combined with jolting makes them frantic. By the time we started, a rich, resonant roar echoed and competed with the sound of the engine. The noise increased gradually, and by the time we had reached home it was perfectly frightful.

We took the box from the car—or rather my friend did—and laid it beside the prepared hive. It was late now, and dark. I was sent for a stable lantern. My friend said it was no use trying to make the bees run in at this time of night; they must be dumped in from the top and then, smoked so that they ran down "into the frames." I said I did not doubt he was right, but he had overlooked one thing, and I suggested he listen to the sound coming from the box. I stipulated that if the box were to be opened I must have a quarter of an hour's previous notice. "What I suggest," I told him, "is that we leave the box where it is until the morning; then, if the bees are in the same frame of mind, put it in a sack with a stone and throw it in the river."

His answer was to put some bits of sacking into a "smoker,"

ignite them, and pump smoke into the box through the hole and the cracks at the bottom.

If the noise had been loud before, it was a mere murmur to what it became now. It seemed as if the volume of it would split the box asunder.

"That's torn it!" I said, but all my friend did was to take the roof off the hive and ask me for a penknife. Why I did not clear off I do not know. I realized this dangerous imbecile wanted to cut the string and let the insects loose. But something held me there: the kind of inertia one gets in nightmares.

"We are now," he said, "ready. Take off that quilt."

The "quilt" was a piece of canvas that fitted over the tops of the frames. I took it off and stepped back—well back; my friend was cutting the string. I closed my eyes and regretted I had lived such a bad life.

There came an increase in the noise; not an increase in the noise made by the bees—that would have been impossible—but the sort of increase you get by taking the silencer from a running engine or opening the door on a roomful of fighting men. Above it my friend's voice: "Don't go to sleep! Raise that light so I can see what I'm doing."

I opened my eyes. He was shaking the box vigorously over the top of the hive, and the bees were tumbling in like currants. It was an anti-climax, but such a welcome one that I was filled with elation. The bees were roaring as hard as ever, but mere noise hurts no one. He banged and bumped till the last bee was out. Then he took the slab of wood on which a number of bees were crawling, lost and disconsolate, and brushed them into the hive with his handkerchief.

I looked inside. A dark mass of insects was sinking slowly into the spaces, between the frames, roaring their indignation, but making no reprisals. My friend expedited their progress with a few well-directed puffs of smoke. Soon the clean white tops of ten wooden frames showed in the light, unspotted by a single bee. We put the quilt back with some warm felt over it and left them. My friend said it was a very little swarm and he must be off. I made him come and have a drink. He did not want one, for it was late and he had to get back. But I insisted. When one meets a hero one always has an urge to stand him drinks, and I classed this man about halfway between a bomb-disposal officer and a lion tamer.

How we change! I now regard my friend as one of the clumsiest beekeepers I know, and his knowledge of bees I class as negligible. He, on his part, despises me. He used to call me a "theoretical" beekeeper (God knows why!), and sneer at the observation hive I kept. He laughed at my worrying about what went on inside apart from the accumulation of honey. He—but I must not run him down. He introduced me to an unknown, amazing world.

I awoke next morning with a vague excitement, which I finally tracked down to the bees. I wanted to go and see what—if anything—was going on. Would they have found the entrance at the bottom? Then I groaned. Of course they would have found it—and taken the fullest advantage of it! I grew annoyed with my friend. Every wild creature has to become accustomed to a new place. And every tame one as well. One does not get a new cat and throw it in a shed and leave the door open. Pigeons, most home-loving of creatures, must be shut up for three weeks before they regard the place as their own. Even men—— Suppose a giant shut up a human family unexpectedly in their house, lifted them up, shook them about, asphyxiated them with smoke, then threw them in a mass into a cellar. Would that family stay there if the cellar door was left open? Or would they bolt like rabbits?

And if *homo sapiens* would behave like this, what could you expect
from insects? No, in his excitement my friend had forgotten to
close the entrance with perforated zinc until the creatures got
used to the hive. By now they were on their way to their old
stand. Or, more probably, just flying about lost, venting their spite
on anything in the vicinity.

Then I remembered I had to see a man about a pony before he
went to market at seven-thirty. It was nearly that now. I jumped
up, dressed, and hurried off. We spoke about the pony. Deals
take a long time in the West Country, and nothing came of it
just then. I told him about my bees and what my friend had
done. He said he did not keep bees, but felt certain they would
have gone. He told me about a cow he had bought that had been
sold ten times before, each time to a buyer at a more distant farm;
and it had always gone back home, however long it had been
shut up. In his case it had swum the river to get away.

As I went back I debated whether to get another swarm or
let the whole thing go. I decided to let the whole thing go. I had
wanted bees to pollinate fruit trees, little thinking it would entail
all this fuss, expense, and bother. By the time I got home I had
decided everything had happened for the best. I had breakfast,
did several things, then went to have a perfunctory look at the
hive.

They had *not* gone. I had got used to that hive as an old dead
box, and the transformation gave me a shock. Lines of traffic
were coming and going. The entrance was not a wooden slit but
the gate of some biblical city thronging with merchants, slaves,
and soldiers. How obvious the soldiers were! How careful whom
they passed! Riches were pouring in, loads of flaming colors,
spicily scented. "Gold, frankincense, and myrrh," I said to myself—quite inconsequently. Could these creatures, so busy, so

My First Bees

ordered in their several duties, be the squirming mass I had seen dumped on the top of the hive only last night?

What struck me very forcibly was that the hive that had been mine yesterday was mine no longer. Even as I watched, four bees came out dragging after them part of a cigarette end I must have dropped inside.

I could not say how long I stayed there watching. I know it was a long time. And to all intents and purposes I stood watching for seven years. The swarm was, as my friend said, a very little one; but it built up. It was a wonder it did. It had much to contend with, myself heading the list. I was forever poking and prying and having a peep inside. Knowing (now) how harmful this is, it amazes me that it survived. I branched out later. I got other colonies and bought expensive queens of guaranteed honey-getting strains, and then queens of almost every known nationality.

Gradually I scrapped them. None could touch the little swarm
I got for nothing. None were so good-tempered or worked so hard.
In bad seasons the swarm gathered twice as much honey as any
of the others. I dethroned the purchased queens and put in their
place daughters of the old original one. And until the war daughters or granddaughters of that first queen that traveled in the little
box ruled over all my colonies.

I went away from that hive a bee enthusiast. The conversion of
Constantine was hardly more startling. A new world had opened
before me. I devoured almost everything that was ever published
about bees. I attended lectures and went in for exams. I bought
special lenses and photographed bees. I bought a microscope and
studied the bee's anatomy. You see, I wanted to know *all* about
bees. My efforts only brought me in the end to that impenetrable
curtain that shuts off from the gaze of man the life of the bee.
He may see so far, but no farther.

CHAPTER TWO

STINGS AND HONEY

This is a book about bees, not about myself, but having embarked on my own experiences, I am tempted to go on a little longer, because the story of my initiation as a beekeeper does not really stop with my first hive. That was responsible for my enthusiasm; my second hive was responsible for my education.

It was autumn, and I was quite happy with my first and only lot of bees, watching them and studying their little ways and making their lives a misery with interference. Then I saw an advertisement in a bee paper: *For sale. Strong Stock of Bees. Miss B———.* The place was only about ten miles away, and I fell to the temptation, though it was really too late in the season to get new stocks. I had an extra hive, ready for a possible swarm next spring, and the idea of seeing this empty box transformed into a busy, self-supporting city like the other was irresistible. Letters were written, the deal closed, and off I went to get my new bees.

Why, I do not know; but for some reason one expects lady beekeepers to be elderly, plain, and a trifle peculiar. Miss B—— gave me a shock by being young and good-looking. I became conscious immediately that I had put on my old "hive" flannel trousers and thought with regret of the suit of natty "plus fours" I might have worn. She was affable enough, however, and talked bees to me for quite a time. "Talking bees," as soon as she learned how long I had been keeping them, meant giving me a lecture on them. Beekeepers are like that—as soon as they get anyone who will listen to them and knows less than they do. That is why I am writing this book. When she had finished her lecture, she said she would go and dress and we could then transfer the bees from the hive to the traveling box. She was a long time dressing, the reason being obvious when she reappeared looking like an Arctic explorer, so swathed was she in overalls, high boots, gauntlets, veils, and other things, all tied up with tapes. No bee could have got at her even if it had studied the problem for a week beforehand. She explained that she was rather susceptible to stings. Apparently she thought other people were not; for in spite of the fact that I had no veil she took it for granted that I was going to help her. She went to the hive she was selling me and opened it, and I stood by and did as she directed. Now her attire was excellent in its complete invulnerability, but it had certain drawbacks: she could hardly see, her movements were hampered, and her thickly gauntleted hands could only manipulate clumsily. She jarred and crushed bees, and bees resent this. Their resentment, perforce, was directed at me.

I like good-looking women and I like helping them, but only within reasonable limits. A Sir Launcelot might have seen it out. In any case he wore armor. Sir John Crompton soon downed the smoker and went to have a look at the poultry. They were black Minorcas or white Leghorns or something. Whatever they were, they had one outstanding quality: they were a long way from the apiary and screened from it by a line of bushes. Even here two bees, sick and tired of attempting the inviolable person of the lady, found me out.

I was coldly received by Miss B—— when I returned, though she relented slightly when she saw my face and inquired with mild interest if "a" bee had stung me. I felt tempted to ask if she thought the rest was some skin disease, but she was telling me what to do with the bees when I got them home, and the kindergarten feeling was on me again. That was my first experience of stinging since I had started beekeeping. It was by no means my last.

My friend helped me to hive the new bees the next day. This friend was the man who had got me my first lot, and he took a proprietary interest in me—more so than I liked. As usual, when he "helped" he did it himself. He was a rough sort of man and liked to impress me with his contempt of bees. He, too, crushed scores and scores against the sides as he transferred the frames, but without the excuse of the lady, for she could not help it and he could. When I expostulated, he said it did not do to be "too sentimental." Anyway, he got the bees into the hive, and they celebrated the occasion by bringing out in a rather pointed manner as he stood there the corpses of the two hundred or so he had killed.

I was, of course, thrilled. It was a large stock, and most of them came out and flew round and round the hive, beginning in small

circles that grew wider and wider and higher and higher. They
knew they were in a strange place, and were orientating the position of their new home. A rather interesting point presents itself
here about moving hives to new localities. You can move a hive
four miles, but you cannot move it two. Within a radius of three
miles you cannot move it more than two feet. The reason is that
the bee knows every inch of the country for two miles from its
hive. Inside that distance it immediately picks up known landmarks, and is off on the old trail to the old home—which is there
no longer. Even three miles is risky; the bee will get into known
country before long and, being an absent-minded creature, may
go to the old site without thinking what it is doing. The bee marks
its home by the locality, not by the hive itself. The entrance, it
knows, is just between those two trees and so far from the ground.
So if you move a hive a foot or so the returning bee flies straight
to where the entrance used to be, and is as bewildered and astonished as if—well, just as bewildered and astonished as *you* would
be if, after your return from the city, you found your house had
been moved a couple of blocks away. For a times it flies aimlessly
about over the stand, but in the end it generally noses out the
hive. If, however, the hive has been moved, say six feet, the
chances of the returning bee finding it are very small. A hive
can be moved from one end of the apiary to the other if one has
the patience to move it two feet only each day (and rainy or
cloudy days do not count). But it is advisable to give the inmates
a reminder. As I have said, the bee is absent-minded. Its thoughts
are running on the harvest and what it will find in the fields.
Nature has not taught it to cope with a house that does not stay
put. So to remind it the beekeeper, metaphorically speaking, ties
a knot in its handkerchief. In place of the knot he substitutes a
heavy rap on the head. He puts a piece of glass slantwise over the

entrance. The bees hurrying off to work in the morning fly straight into it. They soon find the exits at the sides, but the unexpected obstacle causes them to make an investigation before they go off and helps to impress the new position on their minds. A piece of glass ceases to be a reminder as quickly as a perpetual knot in the handkerchief: they get used to it. So if the beekeeper wishes to keep on moving the hive he thinks out other reminders: grass over the entrance, through which they have to crawl—or any dodge that may cause those busy, preoccupied little minds to pause and think.

October passed, and I had to feed my bees and tuck them up for the winter. With much gratification I had seen both stocks slaughter their drones. This showed that they had queens in residence, and I had feared that my friend's unsentimentality might have squashed one or both. The textbook said they must be fed at night and on sugar syrup. I argued that honey, being the bees' natural food, would be better. I also thought they would store it more quickly if I fed it to them in the warmth of day and not the cold of night. My arguments were sound as far as they went, but they did not go far enough. Even sugar syrup coming suddenly out of the blue excites bees; honey has an effect something like neat spirit on confirmed teetotalers. It sends them crazy. Bees never go out at night, so if a can of syrup is put over them in the evening they get excited but remain in the hive, and by morning have had time to settle down and realize that the food is coming from inside the hive and not out, which is a thing that never happens in nature. But I, as I say, put a can of honey over them in the daytime. Some twenty bees came up through the hole in the top, walked down the prepared platform, filled themselves, and went back. Then the fun started. Rendered honey was coming in. Where from? Not from the fields; the bees knew only too well there was nothing to be had there. So the impulsive and excited creatures jumped to the conclusion that the white hive next door was being looted, and the whole lot of them rushed out to help in the good work. But I had also put honey on the white hive, and by now *they* were rushing out to loot the brown hive.

It was my first experience of robbing, and it took me some time to tumble to what was happening. At first I thought the hives were swarming at an unprecedented time of the year, and decided to write to the papers about it. Then I noticed the captains and the shouting, and the fighting and the writhing bodies on the ground. It was most alarming: every second the air was getting thicker with bees, and the noise was terrific. No one can watch without emotion armies of tens of thousands engaged in furious battle. I had that feeling of dithered helplessness that every beekeeper feels to a certain extent when large-scale robbing breaks out. I rushed for a hose, but the trouble was too far advanced by now. The spray of water swept them back in waves, but they came on again with renewed fury whenever the flow stopped, and I rushed for more water. Things looked serious. The stocks might have exterminated each other in their ridiculous endeavor to transpose the honey in the two hives. Then nature came to my assistance— though I did not deserve it. Rain began to fall. Undeterred by my hose, they were definitely discouraged by a downpour that got heavier and heavier, and by the darkness the clouds brought on. Luckily the rain lasted till evening, and luckily, too, they did not resume the fight next day, as some stocks might have done.

That winter, I devoured literature on bees. I got a microscope and studied their anatomy. Not satisfied with English literature on the subject, I perused translations of works in dead and modern languages, all with the rapt absorption of a servant girl over a novelette. The result was that by spring I considered that what I did not know about bees was not worth while bothering about. In short, I had reached that dangerous stage of confidence that comes at a certain period in most walks of life. Boxers, pilots, motorists, and the rest all go through it, and most of them are brought to a right sense of proportion only by some sudden jar. The pilot or motorist in particular is lucky when this dangerous stage of overconfidence does not terminate fatally, to himself or others. What made things worse in my case was that my first stock (the white hive) were exceptional. I could do almost anything with them, and I attributed this to my knowledge of bees and not to their almost freakish amiability. I had not opened the brown hive yet.

The brown hive had been wintered with a crate of honey over the main chamber, and in spring this had to be removed. So one morning I decided to do it. Not having had any dealings with these particular bees since last autumn, I condescended to wear a veil: a flimsy, homemade affair of light netting over my hat tucked in at the neck. I received a few stings straightaway. A more experienced man would have made a mental note that the bees were irritable that morning and taken added precautions. The top crate, I found, had been sealed to the bottom one with comb. Still, it had to be removed. I began to twist and pull.

There was a note in this hive I did not very much like. It had sounded at first like the distant rumbling of thunder—quite a different note from the throaty, almost kittenish purr of the white hive. It still sounded like thunder, but not like distant thunder. The interjoining combs were being broken. In another second I would get the crate off.

If I had only left things now and gone away for a spell it would have been a different story. With the jarring and pulling temporarily at an end, the bees would have got busy licking up the honey oozing from the broken combs; and bees full of honey are apt to take a more indulgent view of life and its trials. But for all my reading I had not learned this. A vague feeling of uneasiness, a doubt that I was quite the expert I thought I was, only made me try to get the job done the more quickly. I wanted to get this over. I pulled and twisted savagely, and at last the crate came up—and so did the bees. They had been waiting to do so.

For one moment the top of the severed half of the hive looked like a pan of treacle boiling over; the next I could not see it for the cloud that was attacking me. I then committed the crowning offense. Holding the thickly populated crate in my hands, I groped my way to the stand placed to receive it, and on the way there tripped and dropped it, causing some twenty thousand additional bees to join my attackers. I did then what I ought to have done long ago, and ran for dear life.

A proper bee veil is so constructed as to hang away from the face and neck. Those who make it construct it in this way with a purpose and not just for fun, as I ought to have known. My homemade veil caressed my face, except at the forehead and, mercifully, the eyes. It was black with bees, all stinging. My hands and arms, too, were black with them. They were up my trousers and sleeves, and those unlucky enough not to find bare skin were doing their business very effectively through flannel and cotton. I ran on and on and gradually outdistanced the cloud toiling behind. Then I stopped and began to get rid of the bees that were on me. They were legion, and most had already done what they had set out to do. Hundreds of little barbs, endowed with malignant life of their own were quivering and twisting in my skin, still injecting poison from the disrupted sac they carried. The owners, their powers of mischief over, part of their bowels torn away, doomed to die a lingering death, abated their fury not one whit. With almost added hate these mutilated creatures were flinging themselves at me again and again, trying to drive in stings that were no longer there. I realized then with awe what the bee is like when roused. I doubt if any male creature could so lose itself in hate and fighting fury. Yet normally the bee, contrary to prevailing opinion, is a timid little thing.

Most people take a morbid interest in the sting of the bee, and I have often been asked what is the feeling and effect of receiving about a hundred stings at once. I can only speak for myself, and people vary a great deal. There are those whom such a number of stings would kill, but there *are* those whom one sting would kill. People "attacked" and killed by swarms die, I think, chiefly from shock and fright. Many animals seem more susceptible than men. Horses suffer intensely. Fowls are killed almost immediately. Other animals such as badgers are unaffected. The bee itself is very susceptible, but has an armor plating penetrable only in the joints of the abdomen and the spirea, or breathing entrances. A fight between two bees is like a fight between two mail-clad knights, each trying with his sword to penetrate a joint in the other's armor. Once a sting penetrates, the bee dies in about one minute in contortions. I once saw a mouse killed by a bee, and it died in about one minute also. Bees hate mice. The smell of one is sufficient for a swarm to leave their chosen abode. This mouse was a field mouse. I disturbed it in the grass near the hives, and it slitherd away and ran into a bee resting on a blade of grass. It was a pathetic scene; the mouse twisted and screamed loudly till it died. It is certainly no wishy-washy stuff, this poison of the bee, and so far its exact formula has defeated our scientists. There are few secrets left these days, but the virus of the bee is one. A wasp sting produces local pain,

The Victim

swelling, and itching. The bee sting affects the heart and circulation. In my case the pain of all those stings soon merged into a swollen feeling. I felt as if I had been blown up with a pump to bursting point. My head felt blown up, too, and throbbed violently, and my brain refused to function. I could not remember where I lived or come to any conclusion on what direction I should take to get home. However, I pulled myself together and started to walk, and then discovered that I had no feet. Either I had no feet or the ground I walked on was not solid—a horrid feeling. Probably a man "paralytic-drunk" feels much the same, and it is as well no one saw me making that uncertain homeward course at that time in the morning. Near the house I lay down beside a flower bed for half an hour.

I *had* hoped to pass it off. I had been putting on airs rather as a beekeeper, and such rough treatment as this seemed ignominious. So I foolishly tried to be evasive before a startled wife. It did not go down. It is no use saying to someone who knows the normal size of your face that one of the bees got a bit cross.

As a matter of fact, the brown hive were a bad lot, and became worse. Their lady owner had told me that she was selling them because she was rather troubled with her heart and thought it would be better to have one hive less to look after. I wonder! I only know her face wore a glow of perfect health rarely to be found in those with groggy hearts. I also know that had I wanted to get rid of a stock of bees I would have selected the identical one she sold to me. But one can hardly suspect a young and pretty woman of duplicity, can one?

From now on the brown hive declared war against me. Then they swarmed. It was my first swarm, and I was delighted—until I tried to hive them. They taught me another lesson then—viz., that swarming bees are *not* always docile and capable of being handled without veil, as the text-books said. Still, I hived them, and to show their independence and their contempt of myself and all my ways they came out again the next day and settled once more on the same bough of the same apple tree. A cautious advance showed that they were still looking for trouble. By now I was getting a little tired of being stung, and I thought of my friend. He was forever coming along and being patronizing and laughing at what I had done. He had opened up the white hive a dozen times, merely, apparently, to sneer at it. I thought it about time he had dealings with the brown hive.

I did, however, warn him about them, but he only smiled and

said one felt like that with one's first swarm. He sent me for a skep, walked jauntily towards them, and when I came back he was running like a stag. As soon as we met he asked me angrily what I had been doing to those bees. I helped him pick out the barbs and denied doing anything to them. He said *some*thing had upset them, and the best thing was to let them stay a little and settle down. He would come, he said, at six o'clock and hive them then. He rang up at six and asked if they were still there. When I said they were, he said he had to see a man about some hay, but would come along later if he possibly could. He never came. The swarm hung there till half past three the next afternoon; then they rose in a cloud and disappeared.

That first season of mine everything went wrong. To explain what went wrong would be to go into technicalities, but it is amazing how many things *can* go wrong in an apiary. In addition, it was the worst honey season on record. I know now I was lucky. Unfortunates who start in a good season get led astray. The whole thing seems so easy, such money for jam. After they have collected their honey they get paper and pencil and do sums. *If two hives bring in two hundred pounds of honey at two shillings per pound, what will be the return on four hives?* This is the first sum. The next one deals with ten hives, or even fifteen. After that the sums increase in difficulty. Here is an example: *If a man buys twenty stocks and half his stocks double themselves every year, assuming that each stock produces each year a hundred pounds of honey at two shillings per pound, what will be the man's income at the end of twenty years?* The answer to this sum makes the beginner who starts in a good year wonder why Rockefeller ever bothered about oil.

As a result of these sums we find a beginner next year possessing a large number of stocks all in brand-new hives. He had done his sums quite well and put down all the items: equipment, shed, etc.

There are, in fact, only two items he has omitted: weather and experience, and it is just these two items that make him do another sum at the end of the next season. The sum goes like this: *If a man spends five hundred pounds in equipping an apiary of thirty hives, what is the gain or loss to the man if no honey is produced, if half his stocks die out through swarming, robbing, or disease, and if the cost of winter feeding for the remainder is at the rate of thirty pounds of sugar per hive at sixpence per pound?* The answer to this sum makes the beginner who starts in a good season wish he had kept locusts instead.

The fact of the matter is that honey-getting is a highly specialized business, utterly different from stockkeeping of any sort. On the face of it, it is marvelous. Your stock feed themselves, clean themselves, and water themselves, and bring to you the produce of your neighbors' lands for sixteen square miles around without tenancy or summons for theft. The trouble is that the bees are doing all this for themselves and not for their "owner." The food they bring in they use for their own purposes. The beekeeper only gets any when there is a superabundance and the bees have to store it away somewhere. Then the beekeeper sneaks in and snaffles it. It does not even pay him to sneak their winter stores. He can feed them back sugar, which is cheaper, but bees cannot rear good young on a diet of pure sugar. A bad season to the beekeeper is often just an ordinary season to the bees. For the beekeeper is only a bee robber. That again sounds simple. But, as any crook will tell you, robbing—if you are to make a business out of it—is very far from simple. An expert thief must sink a lot of money in his schemes, must study the ways of his victims, and even then must be prepared for many disappointments and dead losses.

I was a thief myself, and my honey customers were my fences, the receivers of stolen property. The stolen goods I brought to them

were priceless, but like other fences, they only gave me a negligible
portion of their value. They paid me two shillings for a jar of honey
and then they grumbled. Where else could you buy for two shillings
what has cost twenty thousand laborers and an equal number of
expert chemists and refiners a week's hard work? But I only filched
it myself? I know. But the crook who steals a valuable gem does not
feel he ought to get nothing for it. I had had to lay in an expensive
set of burglar's tools. I had run the risks of my despicable profession
and had been frequently stabbed and wounded. My failures had
been many; the safes I had prized open were too often empty. I
wanted *some*thing. And I wanted something no less because I knew
the goods I brought were priceless.

What *is* this alleged priceless stuff? Beyond a pleasant taste and
smell of flowers what more intrinsic value has honey than a jar of
syrup? To know what honey is one has to fall back on scientific
analysis, and that analysis is not yet complete. Still, the investigation
has been thorough up to date, and we know most things about it.
Honey commences its career as an infinitesimal drop of sucrose
(cane sugar) and water in the flower, called nectar. This nectar
contains other things as well, but the sugar part of it is sucrose. Cane
sugar, however, is not wanted by the bees, and in honey there is
none; it has all been converted by intricate processes, part occurring
inside the body of the bee, into dextrose (grape sugar) and levulose
(fruit sugar). These two sugars constitute roughly seventy per cent
of honey.

So far it is simple. We now come to the other constituents, none
of which you will find in the jar of syrup. First come the essences
of the flower itself, the scented oils and gums which give honey its
flavor and bouquet. Next, less obvious, but greater in bulk and
much more important, is a variety of valuable salts. They are valuable because they are assimilable. The iron, for instance, in any of

the numerous iron tonics sold by chemists probably never gets into the human system at all; the iron in honey does—immediately and direct into the blood stream. These salts include, in addition to iron, phosphorus, manganese, lime, and sulphur. Then we have albumen, fats, waxes, formic and malic acids, nitrogenous pollen, and last, but not least, some very complex digestive enzymes capable of such useful feats as converting starch into malt.

The properties of honey are also varied. Chiefly it is a food; the only food, moreover, that requires no digestion and passes directly into the blood stream. Thus it is often the only food that can be taken by sufferers from severe abdominal complaints. Of its actual food value I think the bee baby itself gives sufficient illustration. On emergence from the egg the larva weighs 0.10 milligram. After being fed for five days on a diet derived from honey and pollen it weighs 150 milligrams. In other words, it has increased its weight fifteen hundred times. The alarming analogy springs to mind that a human baby weighing six pounds at birth would in a few days weigh four tons if fed (and responding) in this manner. Truly the food of the gods. No one, of course, except a fair proprietor, wants a baby weighing four tons. Still, you will appreciate that there must be virtue in this honey. To pass lightly over the rest, it is a stimulant and a tonic. It has a strengthening effect on the heart and is a medicine for the liver. Its acids and salts make it a gentle laxative. It is a skin and hair food. It is an antiseptic also, and when next you cut your finger try applying honey and then a bandage. The speed with which the wound heals will probably surprise you.

If honey were nasty, I believe more people would take it for its medicinal value. As it is, honey is bought almost solely as a sweetmeat, an alternative to jam, and since it is this aspect of honey that appeals to the bulk we may as well study it a little. The delicate flavor and aroma of honey are derived from the flowers: minute quantities of oils and gums in the nectar carried away by the bee. Honey from each separate variety of flower has a different taste. Unfortunately we shall never be able to market these separately flavored honeys. The honey we buy (with the exception of heather honey perhaps) is a blend of a vast number of flowers. That is why there is a sameness about its taste. Honey from one source is extremely difficult to obtain, even for sampling purposes. It is known, however, that honey from lime trees has a minty taste, that from sweet clover (not the common white) a cinnamon flavor; the wild clematis gives us honey with the flavor of butterscotch, and hawthorn, that with a definite taste of nuts. In my hives I used to put glass frames over the tops, and sometimes, under the exigencies of the moment, a tiny bit of comb was built between the glass and the frames. The bit of honey therein was often from one source only; sometimes from a minor, almost unknown source. When I cleared it away I sampled it. The result was a surprising variety of quite different flavors. There was one small lot I tasted that was like old wine. Where it came from I do not know, but anyone could who market honey like that would make a fortune. To mix these superb flavors is really as iniquitous as mixing choice wines of different varieties; but the bees have no option in the matter, and do not care in any case. So we shall have to continue to take our honey as blended by them, and really, on the whole, they do it very well. It is, I find, in spring that one gets the choicest flavors. In late summer I have come upon one or two samples that have been definitely unpleasant. These were probably the product of ragwort and the asters.

There is one other honey, foul beyond words. I met it, of course, my first season—the season when everything went wrong. All the other beekeepers I knew had no honey, but I had some in my two hives, and very proud of the fact I was. There is a tremendous kick in one's first harvest of honey. When you stagger under the weight of a crate that felt like a feather when you put it on the hive you feel as proud as if you had got it from the flowers and distilled it down yourself. I carried the sealed combs to the loft; then with a heated knife (but not with the clean sweep of the expert) I sheered off the outer crust of "cappings." Greenish-black treacle poured out! I remember how I stared and gaped at this evil stuff. It was "honeydew." I have never seen it since, and most beekeepers never see it at all, yet it was ordained that I should get nothing else that first disastrous but instructive season. It took me some time to get over the shock. For long I felt a qualm when the knife sliced away the white cappings of the first comb, and heaved a sigh of relief when rivulets of bright clear amber came flowing out. Honeydew is the sticky substance that appears on the leaves of certain trees, notably oaks, in dry seasons. Only dire necessity will cause bees even to look at it. For them to collect it is as common a practice as for human beings to chew bark. It can never be mistaken. It is smoky, greenish, sulphurous. As far as I know (I have never eaten it and do not intend to), it is harmless. Those who like brimstone and treacle might experiment, and it may well have the same effect.

It may seem unnecessary to touch on such a well-known phenomenon of honey as granulation, but only recently a local shopkeeper told me of a customer who returned six jars of honey saying they could not have been put in airtight bottles because they had all gone "candied." All honey except ling heather honey granulates. It granulates in a longer or shorter time according to its source. Honey from dandelion and charlock, for instance, granulates in about a fortnight. Cold assists granulation, and most honey becomes solid by winter. Those who wish to reliquefy granulated honey can do so by gentle heating in a bath of warm water. Note the word *gentle:* those volatile oils and gums of the flowers are easily distilled away. There is no earthly reason that I can see why anyone should wish to liquefy granulated honey *except* when it has granulated badly; and if it has been in the charge of an efficient producer it will not do so. The palatability of granulated honey depends on the size of the crystals. If these are minute, almost microscopic, the grain is what is called "buttery," and the honey is an improvement, to most palates, in the liquid form. If they are coarse, the honey tastes gritty. Everything depends on the first crystal that forms. This crystal "breeds," and it breeds others of its kind. So all that is necessary for the producer is to have by him (and hang onto it as if it were gold) a supply of granulated honey of the finest, smoothest grain. A small amount of this, introduced in time, "mothers" any quantity of honey. There is one exception. Ling honey, as I said, never granulates, but bell heather honey does, and it granulates into stuff rather like a mass of coffee crystals, and nothing will stop it from doing so. Here again we see the desirability of one-source honey. We would prefer heather honey from ling only, but ling heather and bell heather generally grow together, and the bee is a firm advocate of getting all one can while the getting is good.

CHAPTER THREE

The Outside of the Hive

Not everyone wishes to see a hive of bees opened up, and when the
offer is made most people remember that they are late for some
appointment. But others have heard so much about bees, their
organization, their industry, their foresight, and the rest that they
jump at the chance of seeing for themselves what goes on inside
a hive. Disillusion awaits them. To the uninitiated the city of the
bees is hidden. As the beekeeper opens up the hive and delves into
its heart, bringing out the crowded combs one by one for inspection,
a feeling of disappointment comes over the onlooker. Is this the city
he read so much about—those clusters packed like sardines, those
few bees moving aimlessly about? Even the queen, when pointed
out, is merely a large insect trying desperately to hide itself.

Yes, he is disillusioned. He is also uneasy. The veil provided seems lamentably inadequate when surrounded by myriads of insects whose home the beekeeper is so rashly breaking up. That the beekeeper wears no veil does not reassure him. Beekeepers are "different." He struggles with a desire to flap his hands and run. So he is not in a position to take in soberly and intelligently the explanations given. He goes away gravely doubting the veracity of those who write about bees. He fails to find much difference between bees and flies on a jam jar.

Perhaps, thinking it over, he realizes he expected too much. The bees he saw were not under normal conditions; as well might a giant try to study the inner workings of a human nursing home by tearing off the roof and lifting up the beds with patients, nurses, mothers, infants clinging to them. He may then decide to study bees that are not frightened, undisturbed bees. He finds someone who has an "observation hive"—a glass hive, in other words, where, after lifting curtains, one can see the bees at work. The same disappointment is in store. Masses of insects hang in great clusters, completely inert. Others poke their heads into cells or move about in a vague way. None seems to be doing anything either intelligible or intelligent. In any case, they are all so crowded that obviously no real work can be done. The city is still veiled from him.

In actual fact, of course, the bees were very busy indeed, chiefly on specialized work, while the masses that hung inert were engaged on that most delicate and mathematically exact of all operations— wax making and comb building. Certainly they were too crowded to work properly according to human standards. But bees have too much work to get through to model themselves on human lines. They have learned the art of working in crowds, which means so many more "hands" on each job. But to the ordinary observer who has had no practical dealings with them, bees merely appear futile or inert when they are busiest. Not that even the most experienced beeman sees *very* much. Of their inner councils, if any, of their debates and decisions of their governors and officers, of the supervisors who allot the gangs their tasks (changing them continually so that the nursemaid of yesterday may be the guard of today, while the guard becomes a humble water carrier); of these things he sees nothing.

But, though the uninitiated are out of their depth when studying the interior of the hive, it is not so in the case of the outside. Here, no less than the beekeeper, they can see and understand. It is more intelligible outside, more human-like. Before the aperture of the hive is a long sloping platform called the "alighting board." This is our stage. Here we see the play: those coming and those going, those staggering in with sacks of pollen—the bread of the community—those laden with honey, those bearing water, those carrying, with extreme care, cargoes of the dangerous, sticky propolis—cement for the masons with which they reinforce the flimsy man-made hive and bind it down as if welded. We see the drones, the bucks of the community, swaggering forth, dandified, handsome, brainless like their prototypes, the Corinthians of old, and like them, contemptuous of any honest work. We see the children in their new gray velvet suits brought out by their nurses, sitting in the sun and waving their front legs enthusiastically to the swift, unheeding traffic that passes over them. We see the guards, conspicuous as if arrayed with swords and helmets, at their posts near the entrance, challenging every comer. Yes, it is much more human.

The first tentative question of the onlooker will concern stings. It will probably be put in a facetious way, but this is only a guise. He must be reassured. Probably, like many another layman, he thinks that the one idea of the bee is to buzz up and puncture any human being in the vicinity. If this were the case, of course, there would be no bees. No creature survives—or deserves to—that commits suicide from choice. Bees sting when, in their opinion, a state of emergency has arisen, though it must be admitted that their ideas of what constitutes a state of emergency vary considerably. Each colony is different. In the apiary there will be some that are sweet-tempered and others that are the reverse. There will be some that are nervous, gentle enough until something excites them—thunder perhaps—and changes them from friendly little things to harridans stinging out of sheer hysterics. And there are colonies that are really vicious, that fulfill the dream of the comic supplement artist and hover in stinging clouds about the interloper's head; but they are hard to find. They are almost a freak of nature. No beekeeper "keeps" them, and in the long run their own viciousness exterminates them. The average man greatly exaggerates the possibility of being stung. When a bee comes along just to have a look at him, hovering in front of his face, or when a worker, heavily laden, beating its tired way to the hive, uses his shoulder as a temporary resting place, he jumps to the conclusion that he is about to be attacked. He flaps his hands or waves his handkerchief. This may result in quick, unthinking retaliation on the bee's part, though more probably she will merely buzz angrily at such uncouth behavior and wend her way.

There need never be any doubt. The bee that intends to sting issues no formal declaration of war. She comes like a pellet from a catapult and the sting goes in at the same time. Not that the bee is above a little bluff. When a hive is opened, members of certain colonies make a great point of buzzing furiously around one's face. Anyone not conversant with the habits of the insects would take to his heels at once—which is exactly what they want him to do. Yet actually the last thing these threatening bees intend is to lose their lives by stinging. This is where experience comes in. There is a certain note in the hum of a hive that is going to be dangerous, and when he hears that note the experienced beekeeper goes away and comes back later. Only the novice ever attempts to go against the bees' wishes and "fight" them. And he regrets it. Once really roused, bees are invincible. They could rout (and have routed) an army. Yet treat them in the right way and study their moods and what gentle, friendly little souls they are! Beekeepers are supposed to become immune from bee stings. Some people, in fact, seem to regard beekeepers as a specialized form of life that can be stung continually every day without feeling it—much as the salamander was supposed to be immune from fire. Certainly stings become less painful in time. In fact, six out of every seven may hardly have any effect. Then the seventh comes along, and pains and swells like that inflicted on a mere tyro.

But to harp on this theme is unfair to the bee. The sting she *had* to develop. Otherwise both bees and honey would be unknown. Long ago that tempting prize, that ambrosial sweet that it is the honeybee's misfortune to have to live on, would have led to the destruction of every hive and every bee. Man, animal, bird, and insect join forces in an almost passionate greed for that unique product. We in these times cannot dream what honey meant to our ancestors in the days, not very long ago, when sugar was unknown. Guarded by defenseless bees, the last nest would long ago have been pillaged and destroyed and the honeybee would be catalogued only as a fossil of some dim era. So let us not be too hard on the bee's sting or the dread it fortunately occasions. Though severe and frequently unjust to the bee, when it becomes really necessary Nature often seems to assist her. Here was the bee with a treasure she must defend. So she converted her useless ovipositor into a very useful sting. She could not do it herself; Nature must have done it for her. At every stage in the honeybee's evolution one sees Nature's ultimate collaboration. It is as if Nature were determined she must survive. Certainly without the honeybee and one or two allied forms our earth today would bear a very different appearance. There would be nuts and grass. But practically all our fruits would be absent. To survive the bee has accomplished miracles. She has tackled and overcome problem after problem. At times she has accomplished the seemingly impossible. Yet there have been problems which alone she could not have overcome. More than once, at the last moment, a helping hand has been given to this panting, straining insect in the never ending progress to its dim objective.

Orderly rows of hives, each the same distance from the other, make one rather wonder what this objective is. Up to date it would seem that the bee has worked and schemed merely to provide sweets for an animal already surfeited with sugar. But it must be remembered that she still works as Nature's agent in pollinating her plants and fruit; also that man does *not* "keep" bees. The bees take to those hives of their own accord, run them according to their ideas, and leave them if they feel like it. In any case, man's intrusion into the bees' world is of comparatively recent date. Whether they agree to it or regard it as yet another problem to be tackled we do not know. In a world overrun with men they are probably wise to agree to it. Against man's fires and chemicals, his saws and axes, their sting would be of little avail if they declared war. It was designed for earlier times, when man was a simpler creature and when the bear, the badger, and the wasp were more dangerous enemies.

If honey is a difficult prize to guard it is just as difficult to obtain. Picture a large garden—not a little plot but a place as extensive as the grounds of a manor. Fill its length and breadth with flowers in full bloom, fragrant with scent, and enclose in it a couple of hives. Most people would imagine the inmates of those hives to be living in luxury. Actually, if they could not get away from the garden, they would be dead of starvation in about a month. The bee has to work on a larger scale. She has to think in square miles, not in square yards. Moreover, most of the blossoms in our gardens have nothing whatever to offer her. Weedy patches and unplowed fields interest her far more than the rich, colorful displays that appeal to us. She has no use for the man who wins prizes at flower shows. The bankrupt farmer and the lazy gardener are her friends. It is the little inconspicuous flowers that give honey. Even so, in country rich in neglected gardens and waste fields she must work hard and travel far before she can procure a load worth taking home. The bees get "nectar" from the flowers. Later they must drive off three quarters of its volume to get "honey." Statistics show that the number of miles traveled and the number of flowers visited by bees to obtain one pound of honey verge on the astronomical.

The trouble with flowers, from the bees' point of view, is that they are temperamental. They are not, as seems to be imagined, drinking fountains with nectar always on tap. Conditions must be right before the flower will yield that infinitesimal drop of watery sugar that the bee looks for so assiduously. There must be definite degrees of temperature and humidity. Wet weather is hopeless; overdry weather is as bad. Even the consistency of the soil is of importance; it must contain certain elements. In the course of a week a flower may yield honey only during one half hour. It may blossom and die without yielding once. Even when, to our ideas, the necessary conditions are present, there will be something lacking and the flowers will give nothing. And, on the contrary, when things seem quite wrong, flowers, in their perverse feminine way, may decide to give abundantly.

So far we have painted rather a dismal picture: the toiling bee and the unwilling flower—the tired mendicant knocking at gaily colored doors, curtly refused. It is a true picture, too, but there is a brighter side to it. At certain times of the year Nature spreads lavish feasts for bees. These are called "honey flows" and the time of the year and the kind of crop, of course, vary in different places. Where I was there were two main honey flows, the first in spring, when the fruit trees blossomed, and the second in summer, when the white clover appeared. The fruit blossom "flow" was usually rather a minor affair, yet it was of great importance, for it enabled the bees to build up the strength and numbers necessary to cope with the big event of the year—the white clover "flow," from which they obtained the stores that tided them over the winter. We almost see Nature's helping hand again. There are many clovers; the flamboyant red and the colored mixtures the farmer sows, but all that counts to the bee is the insignificant, inconspicuous, little wild white clover.

But here again conditions must be right. Generous as it is, when in the mood, the white clover will give nothing unless the temperature is about 70 degrees F., and unless the air is humid, the soil moist but not wet—and the rest. But when it does give, it gives in full measure.

It is exciting, this clover "flow." The beekeeper (always on the alert for it) notices the first white clover that appears. This is of no more consequence than a single swallow; it will be weeks before the clover flowers in quantity. Nevertheless, he probably rubs his hands and smiles. There is no telling when the white clover will come. It is an unpunctual flower. It obeys no rules and defeats all prophets. In an early season it may come late; and when other plants are late, it may come early. It may come in early June or it may wait till late July. So unassuming is this little flower that the average man will pass through a pasture full of it and hardly notice it is there. Not so the beekeeper—he will notice nothing else. He will glance at every head to see if there is a bee on it. Probably there will be none. Conditions will not be right.

Then one day—on a hot muggy morning—as the beekeeper stands looking at the outside of his hives, a bee will drop from the sky onto the alighting board. It will be a large bee, larger than usual, and it will drop with quite a thud, like a bit of putty. It will stay where it dropped, panting heavily, apparently unable to move. Then it will crawl slowly into the hive. If the slope is long and steep, it may abandon its efforts to crawl and try to fly, giving little hops, like a wounded duck. While the beekeeper watches it, two more heavy bees fall simultaneously. They also stand panting, then crawl or waddle to the opening. For a period that seems to be the end of it. The fat, grotesque bees have gone inside, and normal bees in a normal manner are going to and fro, as usual, upon their business. Then, in quick succession, two, three, four bees come tumbling down. Out of these a couple fail to make their objective and fall into the long grass. They can be seen later, each climbing laboriously up a blade. Having arrived at a point where the blades sway under their weight, they hold on, panting. Nor do they attempt to move again for anything from five to ten minutes. Then they take to the wing. Their flight has no resemblance to the usual swift progress of the bee. It is more like the floundering first strokes of a would-be swimmer. They barely make the six-inch or so flight to the alighting board. By now bees are dropping in twos and threes, fives and sixes. The alighting board gives out a sound like that of a muffled typewriter. And barely half of them are making it. The grass around is bending under heavy bodies trying to climb. A dense crowd of fat bees is moving slowly along the board towards the aperture. The same thing is happening at the other hives, while from the insides streams of unladen bees are hurtling like grapeshot outward to the fields.

The flow is on. The clover is yielding!

That night the hives will roar like factories in full blast—which is what they are. Battalions of fanners will be massed on one side of the board. From inside, like the bass note of an organ, will come the sound of tens of thousands of wings reducing the nectar to honey. A blast of hot air that would immediately extinguish a candle comes from behind the massed fanners. On the other side, where the space has been left, an equally strong current is being drawn in. The scent of warm honey fills the air.

Minor flows may occur, caused, for example, by lime trees, heather, or areas of sage, mustard, or sainfoin. Such a source may be a considerable distance from the hive—two miles, perhaps, or more. How does the first discoverer tell the others? It used to be thought that she did not, that she had no means of communicating such information and that it was merely a case of one or two of the others noticing a certain bee discharging unusually large quantities of nectar and saying to themselves, as it were, "Hello, what's all this? She's onto something!" and following her when she flew off to find out where she had got it from. These, after their return, were followed by others until the whole hive was in the know. But Von Frisch's studies have now shown that this is not so and that a discoverer *can* tell the other bees that she has found treasure and exactly where it is, also that the other bees do not follow her but go off first, after receiving her directions.

The first thing the forager tells the others is *what* it is she has found. The workers detect the flower scent on the forager's person or in the nectar she delivers to the porters. From this they know the kind of flower that must be sought. It may, for instance, be apple blossom, and if so they will seek apple blossom only, ignoring all other flowers they may come across.

This way of giving information might seem to break down when the honey brought in comes from a scentless flower, but it is useful even then. Very few flowers indeed have no distinctive scent to a bee. There *are* some, however, including runner beans and Virginia creeper, and when a bee brings in honey from such a source (or brings in syrup from the bowl of an experimenter) the bees know that they must search only for a scentless source and will ignore anything that gives off any scent at all.

The bees of the hive are not interested in the find of a forager unless it is from some particularly rich source of food. In the latter case the forager announces the fact by dancing. For ordinary minor supplies of food there is no dancing.

So when a bee dances, the others know that a good supply of food awaits them somewhere, and they know what kind of food it is. They have next to be told how far off it is.

This information is contained in the dance. If the source of food is close by, up to within seventy yards, that is, the bee dances the "round dance," moving round on the alighting board or on one of the combs in more or less a circle. If it is beyond seventy yards from the hive, the bee dances the "wagging dance," the pattern of which is more complicated. Imagine a straight line with a semi-circle on either side: the bee moves from the base to the top of the straight line, turns left, round the semi-circle, up the straight line again, now right, round the other semi-circle, up the straight line, and so on. While she is going along the straight line she waggles her abdomen—hence the name of the dance.

But the bees' foraging range is up to two or three miles from the hive. It is vague and useless for the forager to state that honey is to be found somewhere between seventy yards and two miles. It does not do so. It gives the exact distance (in flying time). This information is given during the run along the straight line and depends on the time taken to make this run. When the source of food is a hundred meters away, the bee traverses the line nearly ten times in a quarter of a minute. As the food gets farther away, the tempo decreases, the straight line being passed six times for five hundred meters, four to five times for a thousand meters, and twice for five thousand meters per quarter minute, with, of course, varying times for distances in between.

The bee's sense of time therefore must be extraordinarily acute, for it does not carry a stop watch—as the human observers did.

Even so, this information in itself is not very helpful. It is no use giving distance without direction. The direction is also given, and with as much precision as if they all carried little compasses and a compass reading were given.

Again it is the wagging dance that is used and it is the direction of the straight line in relation to the sun that gives the bees their bearings. When the dance is made on a vertical comb in the darkness of the hive where the sun cannot be seen, the bee, going up or down the straight line, uses gravity as a comparison, and this has to be transferred by the watching bees to a bearing on the sun. This sounds (and is) complicated. A number of diagrams would be needed to explain it, and it is not my business to go into it here. It is all explained lucidly by Von Frisch himself in his book *The Dancing Bees,* in which he also gives details of his experiments.

In this complicated but marvelously efficient method of conveying information one hardly knows which amongst the bees to admire most, those who give it or those who are able to read it. And although the accomplishment is hereditary, not all bees *can* read it. The young ones are unable to do so at first, as are others amongst the older foragers. Of these latter some continue to go meandering about on their own just as they did before, whilst others get excited and go dashing off, taking no notice of the directions that scores of returning bees are now giving in every part of the hive. There are duffers in every community, and these seem to think they can do quite as well as the others without bothering to try to read complicated instructions. They attempt to enter other hives, presumably thinking a robbing affray has been arranged. Often you may find them buzzing about inside the house. They are looking for the source of the honey that their more intelligent sisters are bringing in.

Unfortunately this enthusiasm over incoming treasure is not confined to what might be termed their legitimate supply. Bees really do not care where honey comes from. During a flow a huge army is kept hard at work gathering nectar. Later, when the flow ceases (and it usually ceases very abruptly on some change of the weather), tens of thousands of laborers are thrown out of work. They become loafers. And out of these (as in human circles) there are individuals who take to a life of crime. Idleness is probably more demoralizing for the bee than for other creatures. She is so eager to work. What is surprising is that more of this vast army of bee unemployed do not take up crime.

A change comes over the hives. The inmates become dour, dissatisfied. They ought not. Their larders are full, the days are still warm. It ought to be the happiest time of their lives; rest after strenuous toil, rest with the knowledge of hard-won treasure safely stored inside. But the bee does not look at it that way. All she thinks about is the stoppage of the supply. Yesterday stores were pouring in. Now all that arrives are one or two laughably light loads a few youngsters bring—from thistles, or something like that. That they have ample food makes not the slightest difference. The bee can never have enough. It is at this time that relations between the beekeeper and his—I won't say "charges," but the occupants of his hives, get a little strained. The bees become irritable. The alighting

board looks like the outside of a labor exchange, thronged with unemployed who get in the way of those who are still trying to do a
little work and of the pollen gatherers and water bearers.

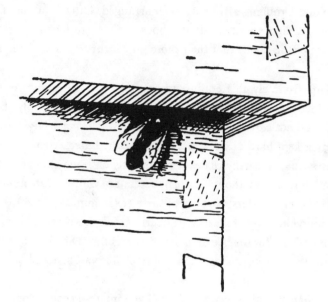

The Robber

But to revert to those few out of the loafers who become criminals—professional robbers, in actual fact. There is no return from
this life of crime. When a bee becomes a robber, a robber it remains.
Never again will it tread the narrow path of righteousness that leads
to the fields and hard work. You can spot these criminals easily. At
our post near the outside of the hive we see the robber bee come
stealing up. His furtive, sneaking manner betrays him. I say "him"
deliberately because it would give a wrong impression to call this
sinister creature "she." The legitimate inhabitants are plainly
marked. They come and go, and if they are intercepted by their

own guards (as often they are) they show all the law-abiding citizen's indignation and by their very demeanor cause the guards to fall back with what look like muttered apologies. Those coming from the fields fly straight up without any deviation from their course. The robber bee sidles round the corner. He buzzes in a shrill, menacing way. He makes a quick dart at the entrance; then sees the guards look up and stiffen at his approach, thinks better of it, and sheers off. He sheers off to the side of the hive, where you will shortly observe him investigating every crevice and trying to squeeze in. Finding no opening there, he tries the back, then the other side. Still balked, he comes to the front again. Though quiet as a mouse when he was looking for unauthorized openings, he again sets up that shrill buzzing as he hovers about the entrance. He does this partly to keep his courage up, and partly, one presumes, to intimidate the guards. He fails entirely in the latter object, for a couple spring at him in mid-air. He eludes them easily (the first thing a robber must learn to do is to dodge) and continues to dart and hover before the entrance. Suddenly, as far from the opening as he can, he alights on the board. The guards, still on the alert for a noisy, threatening robber, do not notice him as he stays quietly there. He now tries to sneak in. Nine times out of ten, however, the guards rush up to scrutinize him, and he springs into the air and discloses himself for what he is.

And so it goes on. Not until he has had several rough-and-tumbles and has been mauled and bitten considerably will he leave that hive and try another. Into some hive, sometime, he will force his way. It may be through a crevice at the back, or he may slip in at the entrance. Once he is inside, his troubles are over for the time being. Provided he is careful those in the interior will not notice him. They are all busy on their various jobs, none of which includes guarding the hive. He must not fill himself up with honey *too* conspicuously—that is all.

He still has to run the gauntlet of the guards on his outward journey. He does not trouble now to waste time by camouflage. He makes a dash for it and, more often than not, comes out the center of a struggling knot of guards. He may be killed. On the other hand, he may get away. He is an expert twister and takes a lot of holding.

This is the average robber. Let us now examine an old hand, a real old lag who has been at the game for ages and has nothing to learn. It is of absorbing interest to watch him. He knows a hundred dodges and has all the effrontery of one of our own confidence tricksters.

He too cannot be mistaken. He is a small, shiny black object. Not a trace of the down that normally covers bees can be seen. It has been all rubbed from him in a thousand fights with guards and by the edges of an equal number of crevices. *He* comes up quietly. The loud, nervous buzzing of beginners is not for him. He alights like a fly on the far end of the board and no one takes the slightest notice. Here he cleans his naked body and doubtless ponders what particular trick in his repertory will be best for this particular hive. Let us suppose it is a hot day. In that case there will be three or four fanners at the aperture busy (in conjunction with others within) at the monotonous job of ventilating the hive. They will be standing motionless, heads down, wings moving so fast they are invisible. The robber's eyes, what time he cleans himself, are entirely on the guards, and when he sees any slight divergence of their attention he moves towards the aperture. By a series of patient, well-timed movements he lands himself behind the rearmost of the fanners.

He has been noticed now. Two guards come bustling up. They

examine in detail what is apparently one of the most conscientious workers in the whole hive. Its head is down and it is fanning for dear life. It is so busy it does not even seem to notice the examination. The guards go away. Both, however, return very quickly when they see the suspect now fanning (equally diligently) in the second row from the aperture. They subject it to an infinitely stricter scrutiny. Still it fans away. One guard goes off satisfied. The other goes off also, but, not quite so satisfied, tries a little trick. As he goes he wheels round like lightning. Actually he wheels a split second too soon. The robber, studying his retreating form, rather prepared for such a move, was on the point of going nearer when the guard turned. He resumes his busy fanning.

Nevertheless, the guard comes back (guards are nothing if not conscientious). Again the searching scrutiny from top to bottom, from antennæ to sting. It is during these prolonged inquiries repeated *ad infinitum* that the nerve of some experienced robber bees breaks. They will stick it out for over an hour, submit to the searches of posse after posse of guards, get nearer and nearer to the aperture, and then, with the game in their hand, spring with a menacing, spiteful buzz into the air.

I loathe robber bees, yet I have groaned aloud when, after an hour's watch, I have seen a jet-black robber lose his nerve like this. Still, the archcriminal will endure all these scrutinies, moving ever imperceptibly towards the entrance, until when the guards come to have another look at him he will not be seen. He is inside.

There are many other roles the hardened criminal will play. He will arrive laden with pollen—sure sign he is merely an honest laborer. He will insert himself into the middle of a little knot of returning workers and march boldly in with them. He will even pretend to be one of the guards and with colossal impudence examine returning workers of the hive he hopes to rob, confusing the guards themselves by his apparent suspicion of others.

It all sounds like much ado about nothing. The robber is only a sneak thief after all. The amount of honey he can filch cannot be large. And he has to work hard enough for it. Unfortunately there is a much more serious side to the matter. Too often robbers are the heralds of war—desperate war, with an appalling death roll. It happens like this. Robbers are continually scouting round the hives trying to get in. Most hives will be strong enough to keep them out, or at any rate prevent all but an occasional thief from entering. But some colonies are weak, or housed in warped, badly fitting hives. It is round about these weak colonies that the robbers congregate. They give them no peace. They shrill about the entrance in such numbers that the guards become bewildered and hardly know which to attack. They penetrate the hive and take their booty home. As they unload they give their message. Others join with the robbers and add to the helplessness of the harassed colony. The colonies of the pilferers wake up to the fact that suddenly, in a time of dearth, rendered honey is being brought in in quantity. It has the same electrifying effect as the cry of "Gold!" in the Yukon in the old days. Honey! goes the shout, and the bees pour forth to get it.

It seems to happen in the twinkling of an eye. At one moment everything is peaceful; the next, the air round the weak hive is blackened with fighting bees. The alighting board is a battlefield deep with combatants. On the grass dying bees by the hundreds writhe in agony. The entrance is jammed with maddened, struggling insects. From inside too comes the muffled roar of battle. The workers coming home with their pathetic little bags of pollen hover, aghast, bewildered at the chaos that receives them. They alight and stand there numbed, not knowing what to do until the sting of an attacker sends them squirming down to join the corpses on the ground.

It is almost impossible to describe to those who have not seen it the abandon of the scene. The savage roar can be heard from an almost incredible distance. The bee—that supposed model of all the virtues—has gone berserk, fighting and stinging like a devil incarnate.

Only night will end the combat (to be renewed next day), unless before then the hive has been wiped out and its stores looted.

And this is not the end of the trouble the robber bee started. If it were it would not be so bad. A weak colony has met the usual fate of the weak. Very hard on them, of course, but—woe to the conquered. What is worse is that the evil spreads. The attackers will not be satisfied with the destruction of one hive. The lust of battle and looting has entered into them. They will never be the same again. Aggression pays. Looting another colony has brought in more honey in a day than they got in a month from flowers. With the same suddenness another hive will be attacked. This time perhaps they will be driven off, but not until the ground is thick with corpses. The spirit of unrest spreads to other hives. It may well be that the whole apiary is soon in an uproar, hive fighting hive.

The beekeeper who notices those first small beginnings of strife and checks them in time certainly saves himself a load of trouble. And it is his duty to keep a careful watch, for he, equally with the robber bee, is responsible. Nature never housed bees in homes each a few feet from the other.

The robber bee does not *ask* for trouble. If he can steal honey from easier sources than occupied hives he is the more pleased. And he has an amazing aptitude for nosing out hidden sweets— and getting at them. I know a beekeeper who invested in a "beeproof" shed and stored his harvest there in the comb—about four hundred pounds of honey. It was beeproof enough for several weeks until a robber got wind of the treasure inside and through some crevice squirmed his way in. The beekeeper returned one afternoon to find his beeproof shed almost hidden in a black cloud of bees. Almost every bee in his apiary was there. A robber had found his way in, and, led by him, the rest of his hive soon found it also. And in a short time every other hive, hearing the uproar, found it too. The shed, when the beekeeper entered, sounded like a powerhouse and was almost solid with bees. Nothing remained of his honey. Where the bees put it was a mystery, but that was their business. Once the game got going, once the apiary as a whole was alive to the fact that four hundred pounds of honey lay in that shed, they found a hundred unsuspected crevices. They got in, but few got out. A crack that will admit a squirming empty bee will not pass a full one. Still, with their tongues those inside passed out the honey to rows of waiting friends outside. It cost the beekeeper some thirty pounds and left him with weak colonies, of little use for the next season—for the mortality was considerable. All on account of a miserable sneak thief of a robber bee.

CHAPTER FOUR

The Tale of a Rebellion

The record of the rocks and the study of allied insects, combined with a little easy guesswork, help us to trace the evolution of the honeybee. It is no dull scientific treatise that unfolds itself, but a drama: a drama of rebellion and civil war, of battle against great odds; a gripping story of the rise to power of a down-trodden band of deformed slaves.

For a starting point we go back countless ages and examine a nest of the original honeybee. It will be in the ground, or the hollow of a tree, or hanging, perhaps, from a branch; any niche, in fact, that may have appealed in some long-distant spring to a single pregnant female bee. It will be small; the huge populations that bees produce now were unknown then.

The female bee has made some rough waxen receptacles and laid eggs in them. The eggs hatch out, and the mother, to the best of her ability, feeds the little grubs. But food is scarce and the weather cold. Undernourished and half chilled, these first children, and several batches that follow, are sadly deformed. They are stunted and incomplete. They are females, but because of malnutrition at a critical period their organs are undeveloped.

Strangely enough, one organ—the brain—has suffered no harm. In this respect they are fully endowed—a phenomenon not infrequently noticed in the case of undernourished human children.

The portion of these monstrosities is unremitting toil, and it is because of their labors that the later arrivals are brought up under better conditions; their food is ample, and the devoted dwarfs keep them warm with their own bodies. Perfect female bees emerge handsome as their mother, full of the zest of life. At about the same time, also brought up by the dwarfs, comes a bevy of young males, great virile creatures.

Life must have been pleasant for these insects then. The days by now were warm and food was ample. Life was a holiday; their only tasks to make love and frolic in the sun and drink their fill of nectar from the flowers.

Meanwhile the deformed sisters—the little Cinderellas of the nest—were growing few in number. Their heavy work had taken toll of them. And such as remained were given no respite—though those they served treated them with contempt. Possibly conditions permitted another batch of young to be raised before autumn put an end to breeding. If so, and if the number of dwarf servants was insufficient, the younger daughters—the perfect females—would condescend to help. Such work was easy now; the days were hot and nectar was abundant. Nor did the thought of food for winter exercise their minds. Only their immediate needs had to be considered. Plenty of time still for sport and love.

The daughters were amorous. The splendor of the males fascinated them, and like other females, they took pride and joy in tending their lovers. They cared for them and even fed them, and soon the lordly male took this as his right. It was degrading for such as he—he came to think—to bother to get nectar from flowers at all. It was a troublesome business. Let the "women" do it. Little did he realize that in this he paved the way for his own downfall.

Then autumn came, and cold nights and mornings began to cramp the ardor of the male—that creature of the sun. The daughters too, all fertile now, forgot their early follies and stayed inside, troubled with a vague unrest. The end was near. These were summer creatures, and summer was over. The drones lingered till they died of want. The females dispersed, each trying to find a snug retreat in which to hibernate. Those who failed to do so died; the others reappeared in spring to found new colonies, in each of which the same sequence of events took place.

And exactly the same thing would be going on today—but for the despised dwarfs.

In dealing with the evolution of the bee we do not pretend to explain how it happens. The evolution of any form of life is mysterious. All we know is that it *does* happen. Without going into the genetics and mutations and Darwinism and Michurinism, roughly, very roughly speaking, what a creature strives for it may attain, after a long period. In the case of the bee a superficial difficulty arises, for the workers themselves do not breed. Yet it is certain that, through the fertile mothers, in some mysterious way the workers evolve. We see it today; not with regard to actual evolution, of course, but with regard to acquired habits. Workers pass down certain traits they have acquired to succeeding workers through a mother who has none of these traits herself and is an entirely different creature.

It has been noted that the little oddities of the nest, though deformed, had their full complement of brain. In addition, they had no sex urges to divert their thoughts or purpose. While their sisters gave their attention to little else but love-making, *their* minds, though unconsciously, were busy in a very different direction. Not long ago, in our own world, a band of poverty-stricken, down-trodden Russians plotted together. We can compare the dwarf bees, at this stage, to that first secret gang of malcontents who finally brought about the Russian Revolution. The task of the bees was harder, and was going to take them infinitely longer, but—probably on that account—it was going to be more durable.

Undoubtedly the first discovery they made was in connection with food. They found that by underfeeding, or withholding certain food of the larvæ, they could, at will, produce others similar to themselves. It seems a retrogressive step, and was probably made, at first, from purely utilitarian motives—to lessen their own labors. It really amounted to an increase in the domestic staff, and doubtless the flighty daughters were in complete agreement with that. Pleasant as their life had been before, it was more so now. No longer need they work at all; the golden age seemed to have arrived. But it was the first and largest nail in the coffin of their freedom. They were in the position of aristocrats on the eve of a revolution.

It was to be a long-drawn-out eve, for the serfs could do nothing yet. Why should they strive and scheme when every winter cut them off, together with the shiftless drones? The only creatures that counted were the princesses, who carried on the race from one season to another. For as yet no honey was stored. There was no object in doing so; the true females would disperse and the others would not live to use it. Nevertheless, the workers, in the evolutionary sense, must have played with the idea for a long time. They found themselves left over in ever growing numbers at the end of the season—a pitiful band, huddled together, dying of cold and starvation. If only they had *food!* There were enough of them to keep warm—if they had the fuel with which to do so.

They soon found out that the cradles once used for the young could be used for storing nectar. And by this means they managed to prolong their lives a few days. But the end was the same. Nectar, as taken from the flowers, is thin, watery stuff. In a week or so it ferments and becomes useless. More than three quarters of its contents of water must be evaporated before it becomes what we know as honey. Moreover, a bee cannot live by honey alone. Honey supplies carbohydrates. Protein is necessary. Pollen supplies this, and pollen also can be packed into the cells. But the same difficulty presented itself. Pollen will not keep. With the damp of winter it becomes a moldy, offensive mass of corruption.

So when the workers began to set their minds to work on the puzzle of living through until the following spring, these two problems were before them. They solved them, and we see in modern hives how they solved them. What we do not realize, but can only guess at, is the patience and tenacity, the study, the trials and retrials that must have occurred before they found the solution. And when they saw the way through they were sadly handicapped by lack of apparatus. In a crowded nest made of wax it is not easy to reduce fifty pounds of nectar to some fifteen pounds of honey, the smallest quantity that will carry a hive of bees through the winter.

They had their wings only for the purpose—small gauzy wings —and few other creatures would have ventured on so heartbreak-

ing a task. By massing themselves in columns they were able to
drive a powerful current of air in any desired direction, one column
fanning fresh air in, another column driving moisture-laden air
out, on the same principle as they ventilate their hive. But to pass
this current over the faces of the stored cells would have been
useless—almost as foolish as trying to make syrup by fanning
sugar and water in a bowl. So they passed the current through rows
of bees, each holding from its mouth the most minute globule
of nectar. After a long period of this treatment it was stored, then
brought out again and the process repeated. And this was continued until the right consistency was reached—and the consistency
must be *just* right.

Even then their labor was in vain until they learned to seal the
honey; for, open to the air, honey absorbs moisture and finally
ferments. The mixture of wax and pollen they had previously
used to seal the grub into its tomb, in which time it wove its shroud
and passed from grub to fully developed insect, was useless here.
The sealing in the one case must be porous, so that the occupant
can breathe; the sealing of honey must be as airtight as a tin cap.
But after all, for a creature like the worker bee, it was a small
matter to learn to make an airtight seal.

The problem of the pollen remained. Sealing was no use—it
went bad just the same. In recent years chemists have discovered
that honey is a powerful antiseptic and preservative. No germ
can live in it. The bees found this out in the dim era of which we
write. They packed the cells only half full of pollen and filled the
remainder with honey. And when they had sealed this honey they
had their preserved pollen—fit to keep for years, if necessary.

With the solving of this problem the situation changed and the
worker bees came into their inheritance—the inheritance their
brains had given them. With their larder full of preserved stores

they could defy their pre-ordained fate and brave the cold of winter.

To recount *all* the difficulties these indomitable little creatures had to surmount would become wearisome. Sufficient to say that this "wintering," as it is called, is a difficult business for the bee. It is a delicate biological operation, and there is every reason to suppose that it is an operation in the course of evolution and not yet perfected. For the bee is almost cold-blooded. It does not hibernate and can live only in a comparatively high temperature. This temperature it has to make for itself during winter. It does so by means of "exercise gangs." These gangs, when the temperature is low, consume a little honey and convert it into heat by violent exercise with their wings, driving the warmed air through the packed cluster of their comrades.

Still, all their difficulties were small compared with those two great problems, by solving which they had bridged the abyss between autumn and spring and gained control of the hive.

Their brothers and sisters did not realize it for some time. The first clash occurred with the princesses—the handsome daughters, who did no work at all now, leaving everything to the little "skivvies." And when these same "skivvies" began to dictate, there must have been ructions. The workers wanted eggs, and plenty of them. They wanted them, moreover, in autumn, when the princesses were thinking of dispersing. And they did not want the princesses to disperse at all. They wanted them to stay so that they could breed next spring. For without laying females their plans would come to nothing. Yes, there must have been lively scenes when the princesses tried to fly away and the workers tried to stop them.

In the fracas most of the queens *would* get away, but some

would be stopped. Later on, no doubt, when custom asserted itself, there would be queens who preferred to remain in the hive rather than face the cold and take their doubtful chances of survival.

It was a breath-taking resolution when, very much later, the workers decided that *one* fully developed female would meet their needs. And they would certainly not have taken this step without making considerable study and experiment in raising bigger and better queens. For it is a far flight from the female bees of the old days in their little nest to the modern queen laying her two thousand eggs a day.

During this period of experiment the idea came to them to abandon the ordinary cell altogether and make a special one. A great affair like an acorn was built in which a large queen could be reared and kept surrounded by a veritable sea of their special queen food, "royal jelly." Naturally, in the trim order of the nest with the growing demand for every inch of space, these cumbersome erections could have no permanent place, and as soon as the queen had hatched they were demolished. It was evidently a successful experiment, for they continue it to this day. The result of their efforts has done them credit. They have evolved an infinitely better egg-laying machine than we have done with our poultry, and they have not ruined the stamina of the race in doing it.

So now we see the flock of handsome sisters reduced to one—and she a prisoner in the darkness of the hive. What of the drones? These in the proper order of things would be the next to be tackled. That clash is taking place now, and so far the workers have by no means had the best of it. Eventually, one presumes, they will emerge victorious. The drone, with his greedy inroads on the food supplies, is too great a stumbling block. But up to date he has

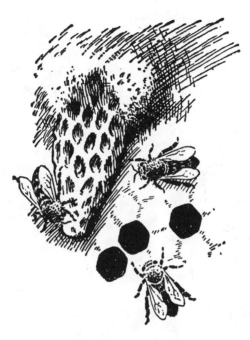

The Queen's Cell

held his own. The yearly massacre that takes place, though doubt-less relieving the feelings of his sisters, does not alter the situation. He lives today as he lived in that first little nest—a creature of sun and warmth—and *free*.

Let us take another look at the drone, for he cannot be quite so stupid as he is made out. The "lazy, yawning drone." He has been vilified since man began to write. But there is nothing lazy or yawning in the way he takes the air on a summer day. With a roar like the engine of a powerful plane he sweeps into the sky. He is never seen again till he returns. No dawdling on flowers for him. Like an eagle, his playground is the heavens. He likes his

comforts, of course, and his food. If the day is cold, he will keep
to the warm hive. But then he is not warm-blooded, as we are.
It is cold up above, and he must spend long hours there. A chilled,
shivering creature would be no mate for a lusty princess. Besides,
no princess either will emerge on a cold day. The workers, at any
rate, do not grudge him his room and meals on chilly days—and
they know.

Late in the summer afternoon the familiar mighty roar announces his return. For all the noise and fuss he makes he alights
softly and gently. He wastes no time. The air above has been fresh
and he has been on the wing for hours. He has, moreover, engaged
in several hectic chases at full speed after various young females—
and "full speed" is not an idle term, for he travels fast. That he
has come back at all shows that he has had no "luck" in his
pursuits. If he had, his disemboweled body would be lying in a
field. He is hungry. He swaggers into the hive and loudly and
insolently demands refreshment. He gets it. He gets it in full measure, without stint, without question, and personally administered.
And, then he goes to sleep.

The drone has the workers in a cleft stick. He encroaches seriously on their precious food supplies, and his numbers are appalling. But they dare not withhold his food or reduce his numbers—
not till summer is over. The whole elaborate system they have built
up depends on a plentiful supply of drones. A princess takes only
one mating flight. There must be no mistake on that flight. Drones
in numbers and at all suitable hours must be quartering the heavens. The princess's rapid flight aloft must not fail to be seen—and
followed. Nor at any time dare they refuse to feed these monsters.
They hope to breed from their queen for years on the strength of
the sperm she derives from the drone on that one flight. What
would be the use of these great egg-laying females they have

learned to breed if mated with a puny, underfed male? No, if they must have drones, as indeed they must, the drones must be good drones; well nourished, virile.

And so the drones represent another problem the workers are trying to solve. What they want, probably, is to get mating to occur within the hive. There need be no inbreeding. Even now there is an intertribal agreement about drones. Whereas workers or queens venturing near another hive are incontinently slaughtered, drones are accepted everywhere. Not only can a drone enter any hive he wishes, but he will be fed there as assiduously as if he were in his own home—and by creatures who, at the bottom, loathe and detest him! Therefore if workers will, for the good of posterity, suffer *any* drones in their hives they could easily see to it that their queen mated with an unrelated partner.

But the drone, so far, refuses to fall in with their ideas. He is the epitome of the die-hard. What he did in the good old days he is going to do now. He demands that his mates be sent out to him. The high heaven must be his bridal chamber. In the hive he will not look at them. Wise drone! Once he *does* begin to look at them in the hive he will be done for.

Needless to say, the drones have not escaped the notice of the beekeeper. He has noted with alarm the number of these hulking brutes and their inroads on the stores. The solution at first seemed easy. He made "drone traps." These trapped the drones but let the smaller workers through. He caught numbers of drones daily, destroyed them, and gleefully computed the amount of honey thereby saved. But the workers would not collaborate. When certain flowers are secreting nectar the frenzied rush of bees to get it is comparable only to the rush men make to a new Klondyke. A miser with his gold is as nothing to a bee with its honey. Yet instead of appreciating the beekeeper's efforts to conserve their hoard they lost heart when their drones were taken from them. They joined but half-heartedly in the rush to the flowers. They dawdled. So the perplexed beekeeper shook his head and left the drones alone. What it amounts to is that no one—certainly not man—can teach the bees their own business.

So that is the position now. The drone has not altered one iota. He still passes the summer hours in luxury and gay adventure. And the omnipotent workers he treats with the same gruff contempt that his ancestors treated that first small batch of "skivvies."

But retribution comes each year. The summer days grow shorter. Unknown to him, his sisters are putting their heads together. The time of mating—the only excuse for his existence—is drawing to a close. Honey still comes in, but not so fast as it did. A slight nip in the air at nights shows that winter is on the way. Anxiously they take stock of the larder. There will be enough, with care, to see them through. Breeding has almost ceased, for larvæ use a lot of honey. What of those great louts that use still more?

The first returning drone comes roaring from above, circles the hive, and lands on the alighting board. He stalks towards the entrance. Inside, he calls peremptorily for food. A diminutive worker is in his lordly way. He brushes her aside. To his amazement she seizes him by a leg. She does it in a timid, half-hearted way and gets flung off as a terrier is flung from a bull. But she comes back, more resolutely now, and another sister nearby helps her, seizing a wing while she tugs at a leg. The drone, a little bewildered, but still intent on his meal, decides to ignore them. He goes into the hive dragging them after him as if they did not exist. Inside, more of the impertinent creatures chivy him. And—incredible thing—none of them offer food! Seething with indignation, he stalks out. Never, he decides, will he enter that hive again. There are plenty of others. He takes to the air, barely noticing that he takes a worker with him,

hanging from his leg and biting at it. He sends her spinning with
what approximates to a snort of anger.

The hive he enters now perhaps receives him. Perhaps not—for
there is a remarkable unanimity in the date hives begin their operations against the drones. If they *do* receive him it will not be for
long. In a few days he will be hustled from every hive he tries to
enter.

The first attempts of the workers to throw out the drone are
merely ludicrous. They never try to sting him, and with his great
strength half a dozen of them are no match for him in a tug-of-war.
They will drag him out a score of times and he will throw one or two
of them off and walk calmly back, hardly realizing that several are
still hanging on. The drone's undoing is the life of luxury he and his
forebears have led; for now he cannot feed himself, and that is
fatal. As his strength declines from lack of food, the perpetual
chivying of the workers begins to terrify him. He and his fellows
cluster, whimpering, * like frightened sheep, in the farthest corners
of the hive—where lie the great walls of honey they are unable
to eat. Even here the workers give them no peace.

He is a pathetic sight when he finally emerges. The workers
ignore him now. Barely able to walk, he moves weakly, slowly,
to the edge of the alighting board. Never again will he zoom
meteorlike into the sky. At the edge of the board he falls, and all
that remains of that super-flying machine crawls slowly, like a
wounded beetle, into the grass to die.

Not *all* drones perish thus. A few—if very lucky—may strike a
queenless colony. The worker bee is not altogether a creature of
instinct. Therefore, at times, it makes mistakes; and in solving some
of its difficulties it has involved itself in others it never contem-

---

* The crying of the drones can often be heard in a hive.

plated. The decision to keep only one breeding queen overcame several thorny problems. And it *seemed* foolproof. A queen is so well protected, so zealously guarded from outside hazards, so strictly scrutinized regarding health and fitness, so quickly replaced if wanting in any particular, that the only danger to her lay in her one short mating flight. Birds might snap her up or she might return to the wrong hive and be killed. Since the latter can only occur in man-made apiaries where similar hives are placed close together, we may exonerate the workers from lack of forethought in this respect. There remain the birds, and here, incidentally, is secondary reason for breeding large numbers of drones. The concourse that follows the princess acts as an unwitting bodyguard. Unlucky, indeed, the queen a bird would take with so many bulkier forms around. And if she *did* meet with disaster there were others to take her place. Did she not return, they had only to liberate the next.

But the best laid schemes of bees, as well as mice and men, gang aft agley. More often than would be expected, the queens sail forth one after the other—and not one returns. That is the end, of course. The colony is doomed.

The workers resign themselves patiently to their fate. They go about their appointed tasks—halfheartedly, but still, they go about them. They keep the hive clean, they bring out the dead, they collect some honey (not pollen; pollen is used chiefly by the young, and they will never rear another youngster.) They also, to a certain extent, repel robber bees; but they have nothing really to fight for, and other hives soon sense the lack of fighting spirit and make a dead set at them. And finally, this queenless colony will harbor all the drones that like to come along—regardless of the time of the year.

It has been suggested that they do this in the pathetic hope that a princess will somehow arise; then they will have drones on hand to fertilize her. This is doubtful. The worker bee is not a fool. More probably the colony merely realizes only too well its hopeless case. The bulk of the stored food was for the grubs next spring. There will be no grubs now. Why go to the trouble of killing off drones? If they want to join the sinking ship, let them!

So the reprieved drones snuggle down with the workers for the winter, and with the workers they die of cold and starvation.

CHAPTER FIVE

The Swarm

In the last chapter I made an attempt to trace the development of the honeybee from remote times to the present day. From a small community which perished in late autumn but sent out a flock of fertilized females—as a thistle sends its flock of parachuted seeds—to hibernate and breed new nests next spring, they developed into the elaborate, self-controlled, permanent society we see today. But in this change-over, marvellous as it was, the reader could not fail to have noticed one glaring omission: they had done away with the automatic reproduction of colonies. They had left themselves with one queen only. They could make others, but the new type of queen they had evolved was helpless. She could not hibernate, or even feed herself. In any case, their present very intricate state could never now emerge from such a crude beginning as a single female bee.

Their society, of course, had become permanent. No longer did the cold of winter strike it down. But no state is immortal. Sooner or later calamity will come. To avoid a gradual process of extinction the bee, if it was to continue on the new lines, had to think out some scheme of spreading itself. It was a problem which no other form of life has ever had to puzzle over. It comes to all others naturally. The plant sheds its seeds, animals have young; ants, wasps, and bumblebees send out their bands of fertile females. The honeybee (by its own machinations) is unique amongst created life in having to increase itself by sober, deliberate thought and preparation. It is unassisted by the urge of sex and unprompted by the natural desire of the young to emigrate—for it is the old, worn queen who leads the colonists and mostly the old bees who follow her.

There must have been many ways of reproducing their societies open to them, but the method they adopted was what we call the swarm. And doubtless they had good reasons, for it is a troublesome and intricate process: and if there had been anything easier open to them, without snags, they would have chosen it. How and when swarming originated no one knows. There have been suggestions and theories, just as there are suggestions and theories about life on Mars and about unsolved murder cases, but actually no one knows or ever will know.

To a casual observer the issue of a swarm of honeybees one bright summer day seems just about as full of glad abandon as anything well can be. They *pour* out, and the air is full of dancing, circling insects whose combined "voices" swell into a loudly echoing song that can be compared to nothing individual. It combines the roaring drinking songs of old days with the richness of a cathedral choir on Easter morning. Amongst the rank and file there *is* abandon. Even the young ones, unable yet to fly, get carried away and many come tumbling out with the swarm. "Tumbling," unfortunately, is the right word, for many tumble to the ground and never get home again. The impression one gets is that the sunshine had suddenly intoxicated the bees in the hive; that they have said, "It's a *lovely* day. To blazes with work! Let's all go out and have a spree!"

Let us examine what plans must be made before a hive of bees can swarm. It will be seen then that there is nothing haphazard about it. Even modern scientific expeditions hardly require more forethought.

The preparations begin some two months before the swarm issues. Those mysterious insects who do the ordering and controlling have taken stock. They have noted the amount of stores and the number of the population; they have made a strict survey of the grubs and nymphs in the cells; for these will be mature, if not elderly, bees when the time for swarming comes. They have studied their queen mother. If she is young and laying well, the casting vote may be given against swarming, and the whole matter dropped for that year. If she is getting on in years, they will consider it a pro-swarming argument. Particularly they will note the dimensions of their hive and the space available. As spring advances the queen must be urged to lay to her full capacity. Is there room for this? If not, another argument in favor of swarming will certainly be put before the meeting of directors.

Let us suppose that at the meeting it has been decided to swarm that season. They must get busy straight away. There is so much in front of them that it will be a couple of months or so before the swarm can leave. First of all comes the building of drone cells. The reigning queen mother will leave with the swarm. A virgin princess will take her place. The princess must be fertilized. Drones must be on hand to do it. Probably there will be drones unlimited in the air from other hives, but the bee is not the type of creature to let matters slide in the hopes that others will attend to them. Besides, at this time of the season these sterile, frigid females rather like drones. They like the feeling of men about the place, hulking, useless louts though they are. The builders get busy. The drone cells are made. Now the queen must be persuaded to lay in them. As a matter of fact she will need no persuading. A queen regards a drone cell as a woman does a new hat: she rushes to it. Probably it is easier to lay drone eggs. Or perhaps she, too, welcomes the chance of breeding sons instead of those never-ending daughters.

The bees must now indulge in a little mathematics. The next item on their program is to raise princesses from the queen mother's eggs. The drone eggs will take three days to hatch, six days to turn into pupæ, and fifteen days to change from pupæ to perfect insects—twenty-four in all; and after this it will be another fourteen days before the drones reach potency. Eggs that are treated to turn into princesses take three, five, and seven days for the same respective transformations—a total of only fifteen days; and the princess will take her wedding flight about six days later. So from the eggs until the two can meet we have twenty-four plus fourteen days for the drone and fifteen plus six days for the princess. Therefore, however much of a hurry they are in, the bees must not start rearing eggs in queen cells until seventeen days have elapsed since the laying of drone eggs.

The rearing of drones does not commit the bees in their own minds to swarming. The rearing of queens does. So when the beekeeper finds those great cumbrous affairs bees use for the rearing of queens he knows that the hive will swarm. Of course, bees frequently "supersede" their queen when she is old, injured, or failing, and for this purpose too they build queen cells, without any intention of swarming. But in this case they use a slightly different

architectural design in the foundations so that the experienced
beeman can always distinguish them.

The weeks go by, the drones have emerged, the time of swarming
draws near. The council decides that, all being well, swarming will
take place in about a week. That day a chosen band flies forth.
They are strong fliers. They leave early and will not get back till
late. Some of them will not get back for several days. They are
called "scouts," but the proper name is house hunters. As in human
circles, theirs is an onerous, thankless job. They fly all over the
country. They investigate every nook and cranny. A load of re-
sponsibility is on them. Generally they do their work well, no
empty hive, no hollow trunk escapes their notice.

In May or June bees will often be seen inspecting any empty hive
in an apiary. They go in and out. They examine the sides and roof.
They examine every part. These are scouts from some hive that
is preparing to swarm. They return and report, and suitable places
they have found will be borne in mind, for the final decision cannot be made just yet.

Everything is ready. It remains to study the weather. The queen
may not have tried her wings for two or three years. Rainy or very
windy weather would be fatal for the far flight they intend her to
take. Continued bad weather would be fatal for themselves also.
They will leave with nothing except full stomachs. They will have
to build a city, rear young, and keep themselves on what they can
get from the fields, and bad weather will mean they can get nothing.
Weather and the queen—these are the two things that now tax
their minds. The weather they cannot control, the queen they get
into flying shape as best they can. Previous to this they have fed
her assiduously, so as to encourage her to lay—babies will be needed
in the hive they are to leave. But that long, swollen queen with her
abdomen full of eggs could not fly a yard, so they take her in hand

and starve her. They starve her till she becomes small and regains
the youthful lines she had before her marriage. No boxer is put into
stricter training as regards diet. There is not a superfluous "ounce"
of flesh on her when the great day comes. Her wings are a little
frayed, but they hope for the best.

Meanwhile the bees may be seen outside the hive scanning
weather conditions as anxiously as human holiday makers in a
doubtful August. If it is going to rain, they will not leave. Let us
assume that it does rain, and that rainy weather continues, so that
day after day the edict goes out: "The swarm remains." It is a
trying time for all. The rank and file, with no responsibilities, get
disgruntled. The work of the hive stops, even during the fine spells
when other hives are bringing in stores. Only a few go to and fro
with nectar, water, or pollen. And as wet day succeeds wet day
things become worse. (If bad weather continues beyond a certain
period, the colony may abandon the project altogether for that season. They then destroy all the queen cells and their occupants.) The
swarm intended to leave well in advance of the hatching of the
princesses, but now these virgins are becoming mature and struggling in their cells. Normally princesses cut a hole for themselves
and bite their way out, but the workers have to assist them by filing
down the apex of the cell so that it is thin enough. Now, since the
swarm and the old queen are still here, they keep them prisoners,
feeding them through a tiny hole they make in the side. But they are
troublesome prisoners. They rage in their cells and shout murder
at each other.

At last the weather changes. A cloudless sun heralds a bright day.
The sullenness of the bees at the entrance has gone. They gather
in little groups as if conversing. Those returning from the fields
forget the cargo they are carrying and stop and swell the groups out-

side. A hum of subdued excitement comes from the interior. The day has come; a vast army is to emigrate.

The youngsters are giving trouble. They rush up and down the combs in between the solid, waiting phalanxes of older disciplined bees. The order is given to load up with stores, and the hum of excitement is plainly heard outside. The members of the expedition fill their honey sacs. They do not fill them *too* full, there is a lot of flying in front of them.

The leader, the queen mother, is excited too. She knows what is coming (better than any of them if she has swarmed before) and is half eager, half afraid. She dodges her escort and like the very young ones runs here and there, badly troubled with "stage fright." Poor queen! For years she has never left the hive or seen the light. Now she must lead a great army into a brilliant sky.

The roar of full-fed bees fills the hive. They are ready, and hang waiting, motionless for the most part. When the order is given apparent confusion will ensue as forty thousand pour through the narrow gate. Already those outside have anticipated the coming order and taken to the air, flying with lightning speed in varying circles. There is a lull of quiet. The alighting board is deserted, the aperture, empty. The order comes. With a furious roar the hordes are released, and a living stream of bees pours forth. Like flood water they emerge in a brown mass. They are carried out rolling, flying, on their backs, on their heads. The air becomes misty, then clouded with bees. A booming, organlike note rises and swells over the fields.

You may witness now for anything from five minutes to a quarter of an hour the complete abandon of an insect holiday. You may watch forty thousand bees indulging in aerial gymnastics and singing as they perform. Where the queen is, no one knows; somewhere among that crazy mob. She came out of the hive in the middle or even the end of the rush. She is in the air now, and there will be those among them who have their eyes on her. She will try her wings, tentatively at first, then take a course according to her strength and temperament. It may be, growing accustomed, she will revel in this new-found liberty. Perhaps she will remember the days when as a svelte young virgin, she shot aloft with a flight so strong that only the swiftest of the drones could overtake her. In such a case the swarm will be long on the wing and will settle eventually in some high tree. Or she may be afraid, or find her body too heavy for her frayed wings, and lumber to a low nearby branch where the swarm will not be long in joining her.

At first the swarm bears no resemblance to a swarm. The air everywhere seems equally full of flying insects. But later, near some branch a small, more compact crowd will be noticed, and later still a lump will appear on the underside of the branch. It is a gradual process, but in ten minutes there is hardly a bee in the air, while from the branch hangs a great pear-shaped cluster. The song and the roar have gone. The holiday is over. Reason and sobriety have returned (if they ever went).

The first settling place of the swarm is always somewhere near the hive. Here they take stock of the situation, make sure their queen is present and unharmed, and that their numbers include the right proportion of old and young. It is from here, too, that they send out the final band of scouts, whose duty it is to choose definitely the new home. They have a number on their list, but some of these may have been occupied in the meantime. ('Desirable residences' are snatched up quickly in the swarming season.) The swarm will stay where it is until the scouts return. The time will vary from half an hour to a couple of days. It depends on the number of places the scouts have to inspect. And while the swarm hangs there the beekeeper, if he wants it, must come and take it.

Everyone knows his method; how he shakes it into a straw skep

and takes it away to a nice shady spot. The most important thing
is the taking away. If he puts the skep near the place where the
swarm settled, the scouts might return before he had hived it; and
off all of them would go. But taking them away to a different place
puts them in a quandary. They cannot leave their queen; nor can
they lead her away when there is no place to lead her to. They certainly cannot afford to take her off on wild-goose chases—she is
much too valuable for that. So they remain where they are, hoping
the scouts will find them.

That evening the beekeeper takes the skep, and with a violent
downward heave shoots the whole lot out onto a long, sloping board
placed in front of the hive that is to receive them. The mass of bees
seems to liquefy, spreads over the board like thick syrup and over-
flows the edges. Abrupt treatment, but necessary. The bees are disorganized. They are scared, and all they want now is to get into
some dark place and safety. Those nearest the aperture see what
promises to be the very thing. They move forward. It is cool and
dark inside, and there is a nice waxy smell. They immediately
acquaint their fellows of the fact. They do it with a special organ.
It is called Nasanoff's organ and is situated at the back, somewhere
about the last ring of the abdomen. It is a white gland, and when
they expose it, it gives off a powerful, aromatic scent—quite a
pleasing scent even to human nostrils. As they expose it they fan
with all their might and send a strong perfumed gust to those behind. It is a signal, and announces that home and refuge has been
found and, furthermore, indicates where that home and refuge is.
On receipt of this news, the queen will often be seen running over
the backs of the bees toward the promised shelter. The swarm takes
the information thankfully but calmly, and becomes a marching
army. Above them, like escorting planes, hover the bees that have
taken to the wing. The forty thousand march on. The aperture is

small and bees are well over the face and sides of the hive. Yet there is no haste, no excitement, no individual pushing.

Discipline has come back. The fanners move in and those behind take on their duties, themselves moving on and leaving the task to those in their immediate rear. So the signal goes on perpetually till the last bee is in. How different is this steady march into the new home from the intoxicated exit from the old! Most decidedly the holiday is over. Already housekeepers are bringing out dirt and bits of refuse.

They are "hived," but it does not follow they will stay. If the place displeases them, or if one of the scouts finds them out, they may all depart the next morning. The beekeeper hived them in the evening, because bees dread the night like nervous children. If night is coming on they will take any sanctuary rather than spend the hours of darkness outside. But as soon as it gets light again, it is a different matter. By morning at any rate they will have made their decision—to stay or not to stay—and if the hive is a nice hive, bees have generally the sense to stay in it, provided, as I have said, that the scouts do not discover them. Actually, the scouts rarely do discover them. They go back to the trysting place to find the swarm gone. They hover a long time as if they thought it might possibly come back, and then they return to the old hive and tell the sad tale of the mysterious disappearance of the expedition and how it never reached its goal. The parent hive, however, is probably much too worried with its own affairs to bother overmuch about sisters whom it never expected to see again in any case. If a sister dies, if forty thousand sisters die—well, they are dead and that is the end of them.

So utterly quiet is the swarm once it has settled that the beekeeper may pass and repass it a dozen times and never realize it is there. In this case the scouts will return and somehow convey to that packed inert mass the news that a home awaits it. * A gradually increasing hissing noise will be heard. This invariably heralds the breaking up of the swarm. Again—in a minute or two—the air becomes full of flying insects. These increase, and the volume of the pendant mass becomes less, and finally disappears. A cloud forms that seems a little uncertain in its movements. It rises in the air, then comes earthward. It rolls like vapor over the grass, goes up or sideways. Tenuous as steam, it goes here and there, forms and re-forms, becomes almost solid, then disperses. It begins to seem that the evolutions will go on endlessly; that the bees are the highly trained members of some ballet chorus, forming and re-forming with faultless precision, so that when they cloud over the top of a high tree one confidently awaits their return. But they do not return. They disappear. Flying swift and straight and sure, they are heading for their new home. Those initial movements, I think, represent the queen getting into her stride, trying her wings for the long final flight before her. Once started, there is no divergence in their course. Led probably by the scout whose choice has been selected, they go straight to their objective. According to the nature of the country, they may fly low or high (a swarm of bees has been encountered half a mile above the earth). The journey will be a fairly long one. The swarm nearly always chooses a distant locality. It seeks to break every bond with its old home; to burn its boats.

---

* Each scout has found a different place, and before the swarm goes off a decision must be reached as to which is likely to be the most suitable. Von Frisch has shown that the information is given by dancing. Each scout dances, but the most vigorous dancer is generally assumed to have found the best place. One by one the other scouts cease to dance and the last dancer wins the argument. The swarm then breaks up and goes to the place this scout has chosen. It may be a short or a long time before a decision is reached.

Bees think. They make plans, and they alter those plans when circumstances alter. We see this at every turn, and we prove it by putting deliberate obstacles in their path just to see how they will surmount them and revise their original plans. All this in the quiet working conditions of the hive. Does reason still function during the madness of the issuing swarm or the torpor of its subsequent packed inertness? Am I right in saying that the swarm is a deliberate, thought-out colonizing effort, or is it some mad instinctive impulse? To those who know or keep bees there is no doubt, but others—who, perhaps, have witnessed the mating flights of ants—are sceptical when they first see swarming bees. Fortunately, proof that everything is well under control is easy. As an example, let us find the queen of any hive likely to swarm, pick her gently up, and clip one wing with a pair of scissors. This means that when she comes out with the swarm she will be unable to fly, though neither she nor the swarm will know it. The swarm issues. The queen tumbles into the grass in front of the hive and is probably lost permanently. The swarm follows the usual procedure and finally settles. Some times elapses, and then the hissing noise announces its departure. But this time there is no drifting about. The swarm goes straight back to the hive it came from. They pour over the front and sides and roof, looking for the queen, and finally go inside. Sometimes (especially when the queen does not leave the hive) the news gets round more quickly. Then the swarm sweeps hither and thither, and at last, without settling at all, goes back to the hive. There are other tests you may make, all of which will show that the swarm is not the mad, delirious, scatterbrained horde it appears.

Generally speaking, the workers themselves choose the settling place. They choose it in advance with an eye to their queen's age and powers of flight, and they try to tempt her there by clustering

themselves. But sometimes she cannot make it and goes somewhere
else, or drifts away and settles on the first spot that offers. This
is what has occurred in those occasional incidents we read about
of a swarm settling on an animal or man (it is called "attacking"
in the newspapers). It may happen also through a queen becoming
exhausted during the second longer flight to the new home.

To go back to the old hive. What remains? On the face of it,
everything; a well-built city, larders crammed to overflowing,
nurseries packed with children (shortly to be useful workers),
and about one-third of the original population of the hive. Everything except a queen. And to provide for this the departed workers
have left a relay of princesses in cells. They are "on velvet," surely.
But sometimes it seems as if the swarm takes with it the brains of
the community—the *crème de la crème* of the hive. Possibly they
think this only right. They bequeath to those they leave their city,
their wealth, their children—everything. They ought not to be
grudged the wisest members—they will *need* them. It is the princesses that cause the trouble. Some remaining colonies are perfectly
well able to deal with them, but others seem to get dithered. The
first princess emerges. Full of youth, impetuosity, and feminine
jealousy, she rushes round and tries to get at the other princesses
still in their cells. If the wise old members of the first swarm had
been there this princess would have been taught to mind her p's
and q's very quickly. She would have been encircled by a guard
and made to do just what she was told—as young females should.
But she cares very little for the half-baked bees that remain. She
replies scornfully to their expostulations and brushes aside the un-
tried guards placed round her. Balked in her attempts on her
sisters' lives and getting more "temperamental" every minute, she
finally sweeps from the hive—and a large number of bees go
with her. None of them quite know what they are doing, but it

develops into a second swarm, called a "cast," and of all the mad gatherings ever let loose on a countryside a cast is the worst. That crazy princess may do anything. Unlike her mother, she can fly with the best. She may settle (not for long) at the top of the highest tree she can find, she may go miles without settling anywhere, or she may go straight into an occupied hive followed by her mad escort. This, of course, will be the proper and fitting end of her and the escort, though unfortunately it will interrupt the work of a conscientious colony, cost them many members, and leave a lot of corpses to be cleared away by sufficiently busy house cleaners.

There *are* occasions, if it is early in the season or if the hive is uncomfortably crowded with emerging young, when a cast— from a colonizing point of view—may be no bad thing. A retreat may be found, and the cast may establish itself. But more often the cast perishes. It makes no plans and takes no precautions, and is just as liable to emerge during rain as during fine weather.

One would expect the issue of a cast to clear the air. The hotheads have gone, surely now the bees will come to their senses. They may. On the other hand they may not. There is no telling what heights of lunacy a hive deprived of its seasoned brains will reach. Normally, when necessity compels, young bees learn sense very quickly—but not in the swarming season. We are still left, you will remember, with a number of princesses in cells hatching out. The next one released also tries vainly to get at the other queen cells, tear them open, and put the occupants to the sting. Finally, just like the first, she too sweeps out of the hive. It is just a tantrum, childish rage because she cannot get her way. The old bees would have taken no notice of it. She could have gone out and cooled her heels and come back when she felt like it. And when she did come back, they would have treated her as if she did not exist—except

to guard her strongly when she showed bloodthirsty intentions
toward the queen cells. They would have let her go to and fro,
but in six days they would have expected her to come back with
her marriage lines, or take the consequences. But once again as
the turbulent princess flings out, a number of bees go with her.
Another cast has emerged. It is suicide and there is no sense in
it. They are dithered; that is the only explanation it is possible to
give.

And it may go on. A third, fourth, fifth, sixth cast may come
out until finally the parent hive, for all the wealth and stores left
to it by the swarm, may be so depleted in numbers that it has no
hope of survival. Soon robber bees will survey the land and nose out
the rottenness of the state. An army of marauders will descend
and folly and inexperience will meet their usual reward.

This is a rather extreme example, perhaps, but it occurs more
often than one would think. It is obvious that the bees' new system
of swarming, though perfectly conceived, does not always function
smoothly. The present era sees bees in a transition stage. They
have evolved a new state, perfect in theory, but a little apt to
break down in practice. There are rough edges still to be rounded
off.

Years ago, writers used to lay particular emphasis on the "docility" of swarming bees. All the old books say that swarming bees
are full of honey and therefore nothing on earth will induce them
to sting. I quite believed this myself until experience taught me
differently. And most modern beekeepers are of the same opinion;
swarms are not particularly amiable. Even normally gentle colonies
become decidedly gingery when swarming. In the old days—the
days of skeps—swarming was encouraged and desired. A hive
which did not swarm got dark looks from the beekeeper, and
probably darker treatment. Now the dark looks are directed on

the hives that *do* swarm, and several complicated processes keep
coming into vogue to stop bees swarming altogether. Some of
them do stop swarming, but they make the bees' lives a misery.
They depend for the most part in mixing the bees up and switch-
ing them round, so that they never know where they are from one
minute to the next. Queens are put in different compartments,
entrances are changed, the nurseries are interfered with at frequent
intervals. All these systems really depend on the bees' determination
to make the best of a bad job and to go on working under any
conditions. (I sometimes wish bees were *not* so long-suffering and
refused to clear up the messes the beekeeper makes, and so teach
him a lesson.) Does this obvious determination of the present-day
beekeeper to stop swarming account for the irritability of the
modern swarm? There is sufficient evidence that the old writers
were right and that in the days of skeps swarming bees were as
harmless as flies, and there is evidence enough that swarms today
are not like that. Do they realize we are fighting their colonizing
system, and resent it? It sounds farfetched, but in dealing with
bees one never knows. They have advanced further than we have
and it is a mistake to underrate their intelligence or perception.
And to fight their swarming system is to fight their independence.
They will never submit to that. Bees and men are a little at variance. The bee thinks she uses man, and man thinks he uses the
bee—and both are right. But man, I think, will find it pays in the
long run to allow "his" bees their annual outing. They will work
the harder.

The swarms, once they have taken over and cleaned out their
new home, work with amazing vigor. For them it is a race with
time. Within two to three months they must build a city of costly
material (twenty pounds of honey goes to the secretion of one
pound of wax). They must gather sufficient honey for their own

Old Ivy

daily needs and for the consumption of the drones and grubs and young—and grubs eat twenty times as much as an adult. In addition to this they must bring in and seal down a six months' supply of honey and pollen for winter, together with a spring reserve for rearing another army of those monstrous feeding babies —call it twenty pounds at least. Only some of them can collect stores. And there are other jobs to occupy the time of quite half their number.

Truly it is a formidable task these colonists have set themselves. No wonder many swarms never see the spring, but die in the winter from starvation. Actually, the zero month is March. This is the month when breeding gets well under way and a new generation is reared out of the sacred winter stores. The bees have no alterna-

tive. They can, and do, stint themselves, but the young cannot
be stinted. The young will take their place quite soon and carry
on. They themselves were mostly bred in the autumn, their dreary
destiny to keep the colony going through the cold of winter and
hand over their job and die when days became pleasant again.
The bees that are doomed through insufficient stores must know
they are doomed. They must know it as early as September. But
they carry on, and it is pleasant to record that in some localities
there comes an eleventh hour reprieve to condemned colonies. It
comes in the rather unexpected shape of ivy—not the trim, well-
kept ivy that clothes the walls of decorous suburban houses, but
ivy that has gone wild; ivy a hundred years old, with roots like a
tree, that strangles oaks and elms and smothers barns and hides
the roofs of derelict outhouses. Such ivy sends out a sea of stalks,
each stalk crowned with a corona of small yellow flowers that
blossom in the unexpected month of November. It is Corn in
Egypt to the bee. Its flowers have none of the fickleness of summer
flowers. They yield their nectar at all temperatures and under
nearly all conditions. So sometimes, in winter, when the hives for
weeks have seemed dead and deserted the old familiar hum that
belongs by right to hot summer days is heard again, and the busy
coming and going is seen on the threshold. And on icy nights the
roar of bee factories, long closed down, sounds once more. To
normal hives this late supply is merely a bonus—superfluous, but
none the less welcome for that. But to those swarms that have
been beaten in their race with time it is the difference between
life and death.

CHAPTER SIX

The Queen

"But mark with royal port and awful mien
Where moves with measured pace, the insect Queen."

So wrote Evans in his epic poem on the honeybee, and indeed
we ought to approach the subject of this chapter with awe, for
no creature has aroused the interest and speculation of mankind
more than the queen of honeybees. From recorded history, and
doubtless long before, men have been intrigued by the mysterious
ruler of a society which itself was mysterious enough. Until comparatively recent times it was thought that this ruler was masculine
and, as all know, Shakespeare's famous lines about the honeybee
make mention of *"His* Majesty," though some of the ancient
Greeks probably knew better; they seem to have got on to most

things by some uncanny intuition. The trouble was that until the moveable frame hive came into use it was impossible to study bees inside their homes. It was clever enough under these circumstances to have found out that they did have a solitary monarch at all, let alone determining the sex. Naturally men assumed this monarch to be male. Not until about a couple of hundred years ago was it realized how completely the insect world is dominated by females. As well as bees, ants were supposed to have kings. Actually the only king in the insect world is the king of the termites, and he is an insignificant nincompoop, a dwarf beside his consort, who spends his time in the royal chamber either prancing about like a goat or hiding under his wife's enormous belly.

It must be admitted that ruthless female dominance has brought the social insects to a high state of efficiency, but let those of our women who would like us to copy them remember that, if we did, women, as well as doing the domineering, would have to do all the work. Personally I would not mind this; I would not mind being dominated if I could put my feet up and just wait for the next meal.

Two things strike one about the queen bee: her length and her elegance. A pair of short wings reaching just below her waist give the effect of a fashionable mantle. From the waist comes the conspicuous trailing abdomen, shining as if burnished, and varying in color from brilliant gold to jet black. Apart from her size you will recognize the queen by her deliberate movements. However crowded the combs, her slow, majestic course is uninterrupted, her way being cleared before her as she goes.

Her daughters and her sons (her very own sons, for they had no father) are sun worshippers. When the sun is hot all come crowding out. The young ones, in their downy gray coats, dance and hover before the hive in a thick cloud, trying their wings.

The foragers, swift and sure, shoot through them like pellets, making for the fields; or, if returning, come sailing up like laden galleons, dropping with audible thuds on to the alighting board. The drones swagger forth from the dark interior, rub their eyes, and clean themselves with a leisurely air of great importance before launching noisily off in search of doubtful adventures.

All seek the sun—except the queen. She must stay in the darkness of the hive. Even a minute—so her daughters think—could not be spared. Too many lives are lost in the hazards of the open fields to permit any stoppage in the queen's output. The teeming strength of the hive must be maintained. Only overwhelming numbers can store those glistening walls of honey—the surplus for the winter and the early spring. They themselves will soon die, worn out by ceaseless work. Others must be on hand to fill the gap, to take over the burden. This is the queen's part. She must lay the eggs, and lay them ceaselessly. Her daughters will attend to the rest. And let it be said that the larva of each of those two thousand eggs a day the queen lays must receive more attention, more nursing, more careful dieting than any human mother could give *her* baby.

Since the queen will certainly not come out we must look for her inside. Only since modern hives were made have we been able to do so. Before this the waxen city was inviolate. Steps on a par with modern warfare had to be taken to storm the citadel: poison gas in the shape of burning sulphur; frightfulness, and the total destruction of the inhabitants. We find her, perhaps, on the middle comb. Solid, piled-up crowds seem to hem her in, but her progress as she moves steadily, almost wearily along is uninterrupted. Those in her direct path seem so engrossed in their various tasks that they realize her presence only when she is almost touching them; then they back away hurriedly, clearing her way. Immediately she has passed they return to their work. On the whole they

appear to take very little notice of her, almost, in fact, to grudge
the slight interruption she caused. Yet actually every bee in that
vast community has, metaphorically speaking, one eye always on
her. Builders, nurses, temperature regulators, brewers, cell cleaners,
undertakers, guards, water carriers, foragers—all, busy as they
are, must keep that one eye on the queen. They like her, of course;
she is their mother. But it is not on this account they watch her;
it is because the life of the colony depends on her. They themselves
are but a temporary output of her stupendous fertility. The bees'
religion is posterity. All labor is for those who come after; all
care for the future of the hive. And they know that if the queen
fails, unless they can raise another within a definite and narrow
limit of time, their colony is doomed. But to the casual observer
they seem far too busy to bother much about her.

Let us make a test. Pick up the queen and take her away. The
workers will not hinder you; the guards are on duty at the entrance,
theirs is the task of repelling invaders. She will not hurt you either.
She has a sting, a curved and murderous weapon; but she is of
blood royal and will use it only on one of her own caste. Now
watch the hive. For some time nothing unusual is seen. The ordinary traffic goes on: the meteorlike exit of outgoing bees, the heavy,
labored arrival of homecomers. Then bees begin to come out who
do not launch themselves into the blue but who run about as if in
search of something. Their number increases. They gather in little
knots as if discussing dreadful news. They rush excitedly at every
laborer returning from the fields, examine her, and leave her abruptly to rush at another. In an hour the outside of the hive is black
with bees: on the front, on the sides, on the roof even, searching
every possible and impossible corner. For two hours this frantic
search goes on; then, one by one, they return. In another hour they

are all inside once more. But it is a different hive. The bustle, the turmoil, the eagerness has gone.

By evening the queenless hive differs in no respect from the others—externally. But experiment again. Go to the other hives and on the side of each rap sharply with your fist. A full-throated roar from the interior is the reply. It is an indignant roar. It threatens reprisals if there is any repetition. It dies down immediately. The inmates are busy. Now rap on the queenless hive. There is a roar too, but a different roar. It is a wail rather than a roar. It threatens no reprisals and it does not die down. For long it continues: the wail of anguish of the lost.

If it is summertime and drones are on the wing, the queenless colony, when its first grief is over, will pull itself together. The situation can be saved. From one of the worker larvæ left by the old queen, a new queen can be raised. Yet no time must be lost, for the larva must be less than three days old. But if it is winter or autumn or early spring, the colony has no hope, their days are numbered.

There is another reason why the workers must keep their queen under observation. They themselves *can* live six months, but in spring and summer they work themselves to death in a few weeks. The queen's span of life is five years. To the workers she must seem ageless, everlasting. But the time does come when she begins to fail. And those daughters who are with her then must be quick to notice it. Whatever their feelings may be for this queen, it is posterity that counts. The mother is failing: too many drone eggs, perhaps, are being laid, showing that the vital fluid she received from her first and only husband is coming to an end. If this goes on there will be insufficient bees next spring to build up the colony after the trying months of winter. (So far must these short-lived creatures think ahead!) The mother must die.

Not immediately. A successor must first be raised. Two or three
larvæ of suitable age are selected, the occupants of surrounding
cells removed and these cells cut down; then the great acornlike
erections for rearing queens are built over the babies; one of whom
—born a mere worker—thus becomes destined for the purple—
the others for the royal sting.

The stage is now set for the murder scene.

The execution takes place at the desire of that mysterious inner
council which first passed sentence of death, and sometimes there
seems a reluctance to perform it; as if the bees put off the dreadful
act as long as possible. But sooner or later the assassins are ready—
about twelve of them.

The queen continues with her work; the ceaseless, monotonous
round. Perhaps she realizes her inadequacy; perhaps, like many
another tired mother, she feels she cannot get about as she used to.
It is with a guilty feeling, doubtless, that she lays drone eggs in
cells made for workers; it is such an effort to make the muscular
contraction which will anoint the passing egg with drone fluid
and convert it from male to female. It may be that she realizes
the well is running dry. But she goes on uncomplainingly, moving
from cell to cell. Then, unexpectedly, her path is barred. The
workers before her do not back hurriedly and clear the way. An
unyielding wall of them confronts her. Behind, and on her sides,
others are drawing in.

No sting is used in the killing of the queen mother. The en-
circling bees fall on her and enclose her in a writhing, struggling
knot. A grim little ball it is, whose fixed, unwavering purpose is
to smother her to death. Blow smoke into it, throw it on the ground;
the executioners merely hug the queen the tighter. The beekeeper
can disintegrate it by dropping it in water, but it would be a
sorely damaged queen he rescued. Normally the assassins will

The Mating Flight

never leave her until the last spark of life has gone. And the queen meets her death with no submission, but fights desperately to the end.

So the new queen is raised and (if things go well) is mated and reigns in her mother's stead until, years later, *her* daughters come to her on the same dark errand.

From this it would appear that there is no sentimentality in the attitude of the workers to the queen. And perhaps there is not. Yet for her they will undergo the greatest of all sacrifices. In confining themselves for sustenance entirely to the products of flowers, bees have placed themselves at the mercy of the weather. Moreover, not being hibernating creatures like most other insects, bees, besides gathering the considerable amount required for present needs, must store up a winter reserve. In poor seasons, when cold, wet, or wind checks the bees' activity or the yield of nectar in the flowers, it may well be that the reserve obtained is insufficient. In this case (unless the beekeeper comes to their assistance) they will die. And the last to die will be the queen; and the last starving daughter will feed to the queen the last of the stores, forgoing it herself.

The highly evolved, and one might almost say artificial, social life of the bee—in particular the relegation of all breeding to one individual—gives beekeepers unusual powers. Like most other things, beekeeping is now on a commercial basis. To succeed, it is necessary to have stocks as nearly perfect as possible. But nothing the ordinary beekeeper can do will prevent haphazard mating of his queens. As everyone knows, the virgin queen, after emerging from her cell, issues forth one fine day and sails into the blue, where she is soon followed by a large concourse of lusty suitors, collected apparently from every quarter of the district. The foremost of this heterogeneous gang wins his bride—and pays a stiff price for her.

Very few observers have actually seen a queen mate with a drone, but several have been within an ace of it. These have usually seen a cometlike procession of drones with a queen as leader. Suddenly the head of the comet falls to the ground. By the time the observers have reached the spot, the mating has been accomplished and the queen gone. Sometimes a few dazed drones are found on the ground, amongst which, of course, will be one that

is disembowelled. Writing for *Nature Magazine* in November,
1957, an observer states how, together with a friend, he saw a
queen mate. He saw a larger insect in the sky drop on a smaller
one. Afterward the two were seen to be queen and drone. He
assumes from this that a queen swooped onto a drone. But in flight
the drone, being much bulkier, appears to be larger than the queen,
especially than a virgin queen, whose abdomen is slender and
almost half the size of that of a queen in full lay. So I suspect he
saw a drone swoop on to a queen and not *vice versa*. Still, his
account is of great interest for he saw the queen fly away after
the two had fallen to the ground and near this place found the
drone crawling weakly about, lacking most of the inside of his
abdomen and near to death. His inside they had seen the queen
trailing behind her as she flew away. They were unable to observe
whether mating occurred in the air or on the ground.

The newly fertile queen is now herself a potential full-sized
colony—in fact, a long succession of them. But every stock is
different. Her marriage—like other marriages—may have been
unfortunate. Her unknown husband (now deceased) may have
come of undesirable strain. His progeny may be poor honey
gatherers, untidy comb builders, lazy, diseased, vicious, inveterate
swarmers, or inveterate robbers of other hives. And the queen will
breed his children all her life.

A stock possessing any of these faults should not be kept. But
if one had to scrap every slightly imperfect stock and buy or raise
another, the business would soon go into liquidation. Stocks
are expensive to buy and very troublesome to move. And they
have to build up. The stock one has, and would like to change, is
a valuable, highly organized, going concern. Infinite work and
patience went to the making of it. Its combs alone, apart from the
magnificent workmanship and the time taken to build them, were

paid for by the bees (and therefore by the beekeeper) at the rate of about ten pounds of honey for one pound of wax. In the cradles are thousands of eggs hatching out under high temperature, and thousands of babies at different stages of development, each receiving specialized feeding and attention. The multitudinous labors of a great city have been apportioned and are working smoothly. Above all, the main army of foragers is daily bringing in wealth. Must all this be sacrificed because, perhaps, the bees are a little gingery?

Thanks to the queen bee and the severely practical lines on which bees have evolved themselves, it is not necessary. The stock can be changed down to the last bee without any interruption of the social system or the tide of wealth. This is accomplished by removing their queen and introducing another. The old stock will rear the new queen's young. Gradually they themselves will die their natural deaths and gradually daughters of another strain will take their place. In two months we shall have our new stock, prosperous and established, and we shall have secured our harvest of honey as well.

It sounds simple. Actually it is not quite so simple as it sounds. There is a snag, and that snag lies in the introduction of the new queen. The bees, it will be found, do not want a new queen; they like their own. The new queen will cost the beekeeper ten shillings or so, and unless he is very careful that ten shillings will be a dead loss within two minutes of putting her in the hive. *And* he will have a ruined, queenless colony—for stocks which have once killed an alien queen rarely take another.

But it can generally be done if certain precautions are taken. First of all (needless to say) the beekeeper orders his queen from a queen breeder. She arrives in due course in a little box with the name and address and the wording in large letters QUEEN BEE

—DELIVER QUICK. Incidentally, this package greatly interests the average postman. Phlegmatic as he is, and accustomed to almost any oddity in the way of letters or parcels, the heading, QUEEN BEE, intrigues him. It is a pity; for he shakes it and listens, and it does not improve queens to be shaken. In this box will be found the new queen, together with from six to twelve attendants who have fed and looked after her on the journey. The box contains two compartments covered over by wire gauze. The queen and her attendants are in one compartment. The other is filled with "candy" (a preparation of sugar, rather like the white inside "chocolate cream")—food for the journey. A small hole, also full of "candy" and covered with cardboard, has been bored through to the center compartment.

Twenty-four hours before this the old queen will have been removed from the hive and any queen cells destroyed. This is zero hour for the bees. They are in the depth of misery at their loss. The new queen is put among them, box and all.

One would have expected them to welcome her with joy—the savior of the situation. They do not. In a dense mass they struggle to get at her—all with one idea, to tear her limb from limb. But the wire gauze protects her (though they sometimes manage to tear a leg off). The bees, you see, must get to know her. More, they must get to like her, and like her so much that they will clean, protect, and caress her—a very far stretch from their present attitude.

Only by eating away the cardboard and then the "candy" in the tube can they get at her. And there is room in the tube for only one bee at a time. The work of getting through takes them several days. When the passage is cleared the crucial moment arrives. It will be one of two things: either she will be accepted and given their homage and love as by right, or she will be slaughtered incontinently. About seventy-five per cent of these introductions are successful.

It is an unnatural process, and it is no wonder the bees are furious when it is attempted. On the other hand, bees have evolved themselves into such calculating, farseeing, provident creatures that, looking at it from another angle, it is strange they do not see immediately the benefit to the colony of this new, miraculously arrived, fertile queen. There is, I think, a touch of atavism in the rage they show to a strange mother; a heritage of the days when bees were family creatures, rather like bumblebees today. Later, reason triumphs.

The fate of the attendants is an unhappy one. It is the easiest way, and the usual one, to leave them in the cage with the queen. In which case for days the hive bees rage at them, and when the terrified creatures are at last released (not so terrified that they have not fed the queen at regular intervals) they are one and all put furiously to death. Better and kinder to take them out and kill them before putting the queen in the hive, and to put in their place, with the queen, a few young, newly emerged bees from the hive she is to occupy. For very young bees are as kind and friendly as little children, and will feed and care for any queen during the time their elder sisters howl and rage outside, lusting for her blood.

The impression may have been gathered that in all these cases the queen is a passive instrument. And it used to be thought so; but she is not—as observation hives have shown. She is sensitive and nervous, changeable and obstinate by turns, typically feminine. Also she is a queen and subject to fits of royal rage. Half the failures of introduction are due not to the hive bees but to the queen. If she is meek and mild, as some queens are, it is almost certain she will be accepted. But as often as not, when the rage of the workers has passed, hers remains. She will have none of these insolent alien creatures. Released, she refuses their homage, dashes aside their caresses, rages through the hive, setting it in a turmoil. There is only one end to arrogance like this. The beekeeper finds her body lying outside the hive the next morning.

On the assumption that the queen is the responsible factor other methods of introduction have arisen, all based on first subjecting her. She is starved and slipped into the hive so that she will beg abjectly for food; or soused in water and dropped among the bees, a miserable, bedraggled object which they will clean and lick shipshape and probably forget to kill afterwards. These methods are generally successful, but the queen is a delicate creature with important functions to perform, and it is questionable if starving or sousing in water does her much good.

Dr. C. G. Butler in England has recently worked on a theory that bees are aware of the presence of their queen only by passing from one to another something they lick from her body, to which he has given the rather cumbersome name, "Queen Substance." Those nearest the queen, he says, lick her and pass this "substance" on to others by feeding them until it has been received by the whole colony. When the queen is missing, they only know it because the flavor of this "substance" is missing in the food some friend gives them. Apart from wondering if the bees in a large and busy hive are feeding each other all the time to that extent this seems to me a very slow and uncertain method of giving vital news. To pass some of the "substance" of the queen on to 50,000 inhabitants must take a long time! And they have to be doing it continually.

The more usual theory is that the queen's presence is known by her scent. I certainly hold this view and I am also sure that when this scent goes, a message is sent by a sort of telepathy to the whole hive. It is known that termites can send "wireless" messages, and I have reason for thinking that bees and wasps can do the same. I once proved this (to my own satisfaction at any rate) quite by chance.

For one reason or another I have often removed a queen from a hive, popping her in a box and taking her away. Nothing happens for about quarter of an hour, then the bees begin to show the usual signs of queenlessness. The reason (I think) they do not know at once that their queen has gone, and broadcast the news, is because the queen's scent remains for some little time on the comb from which she has been taken.

As I say, I consider that I have proved this. I had a fine colony in a fifteen frame hive (containing possibly some 100,000 bees) and I wished to remove the queen in order to mark her. This colony had always been remarkable for its docility and when I took the coverings off and the tops of all the frames lay exposed the bees took no notice and not a one bothered to come and see what was going on. The queen is usually to be found on one of the middle frames, but I generally start my search at one end and go on methodically to the other. I did so now and, almost by a miracle, the queen was on the comb on the extreme right—the first comb I took out. This time I did not pick her off but removed the whole frame, put it in a special case, and in its place substituted temporarily a frame of new foundation wax.

Now the removed frame had the scent of the queen on it, of course; the new one was straight from the manufacturers, and whatever its smell was it was not that of the queen—as those on the frame next to it realized.

The result was amazing. After a slight pause there came a roar from the hive and the bees on the second frame "boiled over" on to the top. In about two seconds the next frame "boiled" also, then the next, then the next, right on to the fifteenth frame, all at about two second intervals. In short, the "boiling" travelled like a wave.

It was not a vague alarm, it was the definite message, *queen missing,* for as they overflowed they started on an obvious queen hunt. So it had taken about thirty seconds to convey to many thousands of bees the information that their queen had gone.

When I had marked the queen I put her back (she, and the bees on the frame with her, being quite unconcerned). I cannot say how long it took for the *queen back* news to get through, but not very long.

By a pure fluke this observation was made in ideal circumstances. First, the colony was an exceptionally placid one; second, the queen was spotted and removed immediately, without causing the slightest disturbance. Normally bees get in a panic when their hive is opened up, and after one has searched frame after frame, backward and forward, for a queen that is dodging about and trying to hide herself the whole colony is disorganized and in no state for any worth-while observation.

S.N. SWAIN after SLADEN

CHAPTER SEVEN

THE BUMBLEBEE

Casual observers would put me down as a great lover of flowers. Seeing me staring fixedly at them for long periods, they would conclude that I was either a botanist, a horticulturalist, or a poet—that is if they did not conclude that I was plain ga-ga. However, except possibly in the last surmise, they would be wrong. I am not a botanist, nor a serious horticulturalist, nor a poet. Confronted with cherry trees in full bloom, for instance, I will stare enthralled, but unlike A. E. Housman, who also stared enthralled no inspiration comes, no lines like his:

> Loveliest of trees, the cherry now
> Is hung with bloom along the bough,
> And stands about the woodland ride,
> Wearing white for Eastertide.

In fact, most of the time I do not see the flowers at all, I only see the bees. The bees, too, I am afraid, have no poetry in them; they are too busy, but they sing, and sing very sweetly, and I cannot even do that.

However, in my early days of beekeeping, this staring achieved *some*thing; it brought me into contact with another insect that I had known before, of course, but to whom I had not been properly introduced—the bumblebee.

When fruit trees and other flowering plants offer their wares, a horde of eager purchasers descend on them like women at a sale. But, unlike women at a sale, there is no fighting or shoving. The customers are bees, and whatever their manners may be at times at home, when gathering nectar they are decorum itself. No bee will deliberately enter a flower occupied by another bee but will search until she finds one vacant. Or she will wait outside until the occupant has done her business and gone away. There are, of course, mistakes; chance collisions, jams, etc., inevitable with such a crowd, but no arguments; indeed the air seems full of apologies; "so sorry," "excuse me," "after you." And the most polite of all are those big furry creatures who might be expected to take advantage of their size and weight—the bumblebees.

Naturally, I wished to find out more about these co-workers with my bees. What did they do with themselves when not getting honey? How did they conduct their domestic affairs? What size were their families? But they cannot be studied like bees in a hive; only by breaking into their underground nests can one see them at home, and to do that is to condemn them to a lingering death without being much the wiser. The only way is to put a colony into an artificial nest, but usually they will not collaborate and will fly away and perish. Luckily a great man named Sladen had been over this ground before. In his classic, *The Humble Bee,* he describes his many unsuccessful efforts to house bumblebees until at last he hit upon the only (but by no means certain) way. It is no use putting a fertile queen into an artificial nest, nor a queen with eggs, nor one with larvæ; she is terrified and as soon as she is allowed her liberty rushes away never to return. Nor can one just dig out a nest and transfer it. The only way is to wait until a nest is fairly populous and then, one by one, catch each individual member including the queen and put them in a bottle. After this the combs must be dug out and placed in a prepared nest, which has to be as like a natural one as possible, and the bees released into it from the bottle. Then, perhaps, they will stay. Even so, they are not in a very convenient place to study and it all has to be done afresh every year.

Catching each bee separately and putting it in a bottle might seem rather a dangerous operation, but the bumblebee is so gentle and good-natured that it is very rare to receive a sting. I have done it a few times and out of hundreds of bees only received one sting, and then because I happened to be dealing with *Bombus terrestris,* a species inclined to be more assertive than most. One thing about them helps the investigator: although they live normally in the dark, they do not seem to be affected by light. Provided their prepared nest is opened without causing disturbance or vibration they can be seen acting in a normal manner.

Hoffer is another authority on bumblebees, but Sladen is the better known. He spent his life, including his boyhood, studying them. His book may well live forever, but I found out only recently how quickly a writer can be lost. A few years ago I received a small package from Canada. Inside was a little book, homemade out of parchment paper neatly sewn together. The text was carefully hand printed, and the book was full of beautiful and accurate paintings of bumblebees together with their scientific names and a full account of their lives and habits. It was done by F. W. Sladen.

The covering letter was from a lady in Owen Sound, Ontario, who said the book had been given to her by an elderly person who had asked her to deliver the book to F. W. Sladen. It had been made by him, she had added, when he was fourteen, and she thought he ought to have it. Would I please, said the writer, give this book to Mr. Sladen.

Now I knew that Sladen had written other books and treatises including in 1902, an important one on the scent organs of honeybees, had been a lecturer, and a member of several bee societies. He was also a household word amongst beekeepers on account of the many hive gadgets he had invented which are still in use and sold under his name. So I anticipated no difficulty in finding out about him and wrote to a bee society for information. They replied that he had been an active member, but they did not quite know where he was now. I wrote to scores of people after that and the result was a blank. All had known him, but I could get no particulars whatever about him. I could not even find out if he had a family or how old he was, though one correspondent did add that he thought Sladen must now be nearer one hundred than ninety, which made my chance of giving him his book seem remote. In the end I presented it to the Natural History Museum, where it is now.

There is a postscript: a year later one of the correspondents told me he had been informed by an acquaintance that F. W. Sladen had been drowned in Canada while duck shooting in 1917.

But it is time that I returned to the subject of this chapter.

Sometimes on a bright day in early March a distant, resonant song may be heard. It grows nearer and louder, and then a bumblebee in its gaily colored overcoat swings out of the sky. The scene changes; the day seems brighter, the sun warmer; if a bumblebee comes, can spring be far behind? Unfortunately it can, and before the sun has set that bumblebee will probably be hunting for some sheltered niche and wondering why it ever left the old one. But the song of the first bumblebee is an overture to the chorus of honeybee and bumblebee voices that will sound later from the fruit trees.

The chief difference, of course, between these two is that bumblebees are annuals; when the honeybees are beginning to think of tucking themselves up nicely for the winter, the few remaining members of the bumblebee families lie stiff and cold, close to the large form of their dead queen. We *call* her a queen, but she is not; she is the hard working mother of a large family, who never ceases slaving from dawn to dusk or dusk to dawn. The queen of the honeybees *is* a queen—at least she fulfils the fairy tale conception of such; never does a stroke of work, has all her actions and her every step mapped out, and is served by a retinue of attendants who do not even allow her to wash or feed herself or to have anything to do with her own young children. And that, I suppose, is how a queen should be. The bumblebee mother cuddles her first children with an almost passionate maternal love, and as they grow up, never ceases to marvel at their attainments. At least, that is the impression she gives. But then, alone and unaided, she has had to see to the making of the home, the laying of the eggs, the incubating, the feeding and warming of the larvae, the laying in of extra food, and a thousand other things including protecting them from enemies. They are her own babies, the fruit of her own efforts.

Before examining the family life of the bumblebee, I should like to make a few general remarks, and if these are already common knowledge I can only apologize and point out that this is not so among my own acquaintances, one of whom suggested that I wrote about humble bees as well as bumblebees—but doubtless I move in very ignorant circles. The only course seems just to go ahead and ignore the long-suffering looks of readers who knew it all before.

There are many species of bumble (or humble) bees and most of them are pleasing in appearance. Bands of gold, yellow, white, tawny, or red on a black or brown ground are their favorite colors, though *Bombus ruderatus* generally (not always) dresses in black. "*Bombus*" (which the Latin dictionary tells us means "a deep sound," "a buzzing") is the name for all of them—*Bombus* this and *Bombus* that; and on the whole systematists have acted with restraint in their second names, though *latreillellus* is rather a tongue twister. Unfortunately, so far as I know, none of them possess nicknames, which is surprising for such familiar creatures.

Bumblebees are choosy about their habitats and will be common in one spot, for no apparent reason, and rare in another close by. Their terrain, too, differs; the beautiful *lapponicus,* for instance, insists upon mountains while the bad-tempered *muscorum* wants marshes. As a race they are inhabitants of the north and, unlike most insects, hate hot countries. A few species are even found in Greenland and Alaska, but none in India or Africa unless it be high up in the mountains. Nor did any exist in Australia or New Zealand until some were sent there.

Naturally, with this preference for cold countries, bumblebees are hardy. Unlike the honeybee they begin work in the fields when very young, and what is more start work earlier in the day and finish later. This does not mean that the honeybee is less energetic, but it is certainly softer. In the early morning it pokes its head out of the hive, shudders, says "no thanks," and goes back and waits until the chill is off the day—and personally I do not blame it. The bumblebee shoots unquestioningly out of the nest as soon as it is light.

The tongue of the bumblebee is longer than the tongue of the honeybee (that of the very abundant B. *hortorum* is almost the length of its whole body) so it can tap honey sources such as honeysuckle and red clover, denied to its more advanced relative. It follows therefore that we owe red clover almost entirely to the bumblebee, for the honeybee finds it extremely difficult to reach the nectaries and effect pollination. And this is a pity, for the red clover produces more honey than most other flowers. Children, cunning brats, know this—or used to—and pull the petals off to suck the sweetness at the base.

Red clover is valuable for stock-raisers, and Australia and New Zealand needed it, so in 1884 bumblebee queens were sent out and became established, and now these countries can use their own red clover seeds. The queens sent out were of two species, B. *terrestris* (the commonest variety) and B. *ruderatus* (the black lady). It is a peculiar thing, but most of the live stuff sent to Australia seems to go wrong. Consider rabbits, which nearly ate the country up, and prickly pears, introduced from South America, which threatened to turn the whole of Australia into an impenetrable thorny thicket until a certain small brown moth was brought along to save the situation. I will say at once that the black queen behaved in an exemplary way and has done ever since, but *terrestris* did not. This bee in several minor respects is not always as virtuous as she might be, and after she had been in Australia some time she began to wonder why she bothered to force her way into foxgloves or snapdragons or other tantalizing, long-petaled flowers to get the honey.

Surely there must be some easier way. She found one. She cut a neat hole in the base of one of the petals, poked her tongue through, and got the nectar that way. Which was cheating. And what was worse, the flower was ravished without being fertilized.

The black queen, as I said, was above this sort of thing, but she was not above using the short cut already provided by her fellow immigrant; nor were the honeybees. The Australians did not worry about foxgloves or snapdragons, but they did not like their broad beans being treated in this way and the crop halved. So they wish they had been sent some bumble less ingenious than *terrestris*. But it is too late; like the rabbits, *terrestris* is there now.

As I have said, the bumblebee is loath to use her sting. The exception is *B. muscorum,* though our friend *B. terrestris* is inclined to be temperamental at times. *Muscorum* is a brown and bright yellow creature and decidedly hot. Take no liberties with her. To make things difficult, she is practically indistinguishable from *helferanus,* who, in spite of her name, is the meekest and mildest of all the bumblebees. Another virtue of the bumble is that when it does use its sting the punishment is not half so great as that from the sting of the honeybee. I hardly need say that the males do not sting, and you can tell a male from a worker by counting the segments of its abdomen: it has six instead of the worker's five.

With us it is usually the females who indulge in scents (a trait supposed by some anthropologists to be an atavistic legacy), with bumblebees it is the males. With one exception they exude a sweet and most delightful perfume. When they leave the nests, which they do as soon as they can fly, they have certain spots which they visit frequently; little holes in a bank, cavities in tree trunks, and similar places. And at these spots their perfume remains. The exception is the species already mentioned with the long name—*B. latreillellus.* These bees possess a disagreeable smell, rather like rotten hay. It is

not only the males; the queens and workers have it too, and so does
the nest and comb.

And now for the short and simple annals of this homely bee. We
will start with the queen that came along singing that bright March
day. She has come from some little hole in a bank or the thatch of
a house or some similar place, and the sun has made her think, like
us, that spring is really here. She will discover her error, as I said,
before the sun has set and will probably find another retreat. But
sooner or later she will have to come out in earnest and start house
hunting. Exactly when she does that depends on her species; there
are early and late species, and those that start early close down
early. *B. pratorum* is the earliest of all and she may make her nest
in March, while late species like the mild *helferanus* will not start
till June. But they will all come out on warm days and visit the
peach blossom or the willow catkins or whatever offers.

On some bank, probably, the queen will find a place that suits
her—often an old mouse hole. With some exceptions she will select
one approached by a longish tunnel; about two feet is the average,
though seven is not unknown. Inside the cavity she makes a cosy
nest of grass or similar stuff, and on the floor, in the center, she
builds a bin. It is a rough, more or less circular affair made of soft,
dirty brown wax. Those accustomed to the precision, superb craft-
manship, and white beauty of the honeybees' comb may smile at
the crude efforts of these rustic folk. Everything they make is rough
and ready. But then, is it fair to compare the product of a single
female, who is not only pregnant but has all the work of the house
to do as well, with the achievement of a specialized gang working
on nothing else? Anyway, when she has made her bin and put some
pollen in it she gets to work on a honey tub, which is, if anything,
cruder than the pollen bin. Then (sometimes before making the

honey tub), she lays her eggs, seven to fifteen of them, in the pollen
bin and seals the bunch over with wax. She now sits on this—what
we will politely call—cell, day and night, except for occasional ex-
cursions to get food. It is really quite remarkable to see the normally
so active bumblebee squatting motionless like a broody hen on her
clutch. Her foraging expeditions must not take her away from her
eggs long enough to chill them, but even so she manages to lay in
about half a thimbleful of honey in the tub. It is thin, watery stuff,
just untreated fluid from the flowers and liable to ferment, but what
do you want? Do you expect her to render it down all by herself and
collect it and hatch out eggs as well? And fight also? For even her
pathetic hoard of honey and her eggs and grubs and cocoons are
looked on avariciously by a crowd of creatures from mice to ants.
So much so that of all the queens that set up homes in the spring,
only a few survive.

In four days the eggs hatch out and the tiny larvæ begin eating
the pollen that forms their bed. They are also fed by their mother,
who makes a small opening in the wax covering and gives them
doses of a mixture of pollen and what she calls honey. In a week
(eleven days after the eggs were laid) the little grubs are full grown
and each spins round itself a cocoon. They can be seen now; for as
soon as they have completed their shrouds the mother takes off the
wax covering and the pale yellow chrysalises are exposed. They still
have to be sat on. In fact, the mother has another fortnight to go,
and she becomes more like an old hen than ever, trying to flatten
herself out in her efforts to cover every one. And all the time she has
to make her pilgrimages to get food to keep herself alive and to set
by a store which very shortly will be badly needed. On that honey
tub now and its contents depends the life of this new bumble family.

In eleven days, twenty-two days after the eggs were laid, the first
daughter bites a hole in the top of her cocoon and struggles out. I

have not really got accustomed to this conjuring trick of Nature's yet. Figuratively speaking, I scratch my head every time. "Here," says Nature, "I have a maggot." And then—there is no deception —out of the silk hat with which Nature has covered the maggot emerges something that in every conceivable particular is entirely different. None of the Biblical miracles were half as good as this one. What emerges from the first cocoon is not quite a bumblebee yet. It is a grayish, silvery thing with a matted, glued-down coat like that of a new chicken. The mother, thrilled to the core, lets it get from under her and watches it anxiously as it totters to the honey tub. There it clumsily unfolds its brand-new, neatly rolled proboscis and takes a drink; then staggers back and quickly nestles again under mother's breast.

One by one the others come out, those at the edge, where they have not received all the warmth, being the last to appear. It may be a month from the first egg before they have all emerged. It takes two days for these new-born bees to get a fur coat like their mother's, though of course they will never be the handsome creature she was; they are only half her size, and their little coats look more like cheap imitations than the real thing. In three days they are flying to and fro and bringing in pollen and honey. The nest has started. The girls are working.

Truth to tell they have not come any too soon. The continuous and increasing work has told heavily on their parent. She has made the nest and brought them up, and she has paid for it in the way hard-working women do. She has lost her looks. Her once handsome coat is frayed and patchy, her wings are torn, and her figure has gone. She bustles about a lot and still tries to do everything herself, but she has nothing like the energy she had. Luckily her daughters take her in hand. "Now you leave this to us, Ma," you can almost hear them saying. "You've done quite enough."

And gradually, as the daughters get stronger and more experienced and can bring in enough honey and pollen, the old matron confines herself to indoor duties. She lays other batches of eggs. She insists on helping to incubate them, but her willing daughters can do it just as well, and they can feed the young, too. So more workers are reared and the family grows larger. These younger ones are bigger and stronger. The mother did her best with her first, but they were not so warmly kept nor so well fed as when she had assistance.

I expect there is a deal of joking at mother's honey tub. Even the first lot soon refuse to use it, and make others, none of which are any better so far as I can see, but mother's tub has certainly got moldy inside. They also put honey in the vacated cocoons, which they patch up with their brown soft wax that reminds one of Plasticine. Yes, the nest is getting busy and domestic now, with a place for everything—more or less.

Unlike the honeybees, the workers bring in honey and pollen at the same time. First, they unload the pollen from their baskets into a bin, then they disgorge their honey into a tub, then they are off again for another load, not even taking the half hour or so rest of the honeybee. As with the latter, night brings no respite; building, incubating, feeding; work is the order of the night as well as of the day. Time presses; the life of the bumblebee nest is but three months.

The rearing of workers comes to an end. The nest is at its prime. Its inhabitants number—well, it varies, one hundred to two hundred will be the average, though there may be more, and it is filled to capacity with stores. How much honey altogether? You who are beekeepers, thinking in terms of a hundred pounds or so per colony, do not smile—one or two ounces is the answer. All is set for the production of sons and fully-sexed daughters. First appear the

sons. After the great hulking louts the honeybees breed, the males
of the bumblebees come as a surprise. In most of the species they are
indistinguishable from the workers in size or appearance, and in
morals they put the drone to shame—or would do so if the drone
had any shame. That ne'er-do-well is an encumbrance from the
day he is born. He loafs around, refusing to go out at all unless the
weather is perfect, and greedily consuming large quantities of the
laboriously gathered stores—which, mind you, have to be personally administered to him. The bumblebee male refuses to be
a burden to his mother and sisters. He says good-by to them as
soon as he can fly and goes off into the outside world to earn his
living—and he does not come back. Doubtless they miss their men
folk, but this poor working-class family could never afford to keep
them in idleness on their slender means. So with the bumblebees we
are spared the harrowing sight of sisters chivying their brothers to
their deaths; though somehow I feel that the mild and gentle
bumblebee never could do this; that she would rather starve. The
sweetly scented males now live a wandering life in the open and get
their own food during their short three or four weeks of life, and
before they die, on their journeys, they meet princesses and perform
the duties for which they were born.

Now come the last-born, the perfect females; great, strapping,
wenches as handsome as their mother was—but is no longer. These
stay at home awhile and are not above helping with the housework,
bringing in loads of pollen when they manage to remember. But
their minds are on other matters, as with most young females, and
as is very right and proper in due course, they, too, take their departure and go about with some boy friend.

There is no hard and fast rule about the breeding of these males
and true females. System and method have little place in the
bumblebee home. The females may be raised before the males, or

no males may be raised at all, or there may be males and no females. But usually the males come first, and their number is roughly double the number of females—say a hundred or more; it all depends on the domestic staff available.

All this time, from their birth, the elder daughters have worked like slaves. The work has been shared, without any method but with complete harmony. There has been no shirking of the more distasteful or laborious jobs, and the mother too, in spite of her failing strength, has tried to do her bit in everything except the outside foraging. Her daughters have paid her no sort of homage and no exaggerated respect, nor have they attended to her personally. They have all mugged along amiably together and never has there been the slightest discord—until the laying of the male and female eggs. Then there are little gatherings and mutterings. Something like this happens often enough in human families; sisters become jealous of younger sisters that are more fortunately endowed or given advantages they never possessed. And the bumblebee family is in more ways than one like a human family. Somehow the elder sisters know that these eggs are destined to become females with opportunities and careers far removed from theirs; females who will each have a mate and who will not even be asked to do their share of the daily work. With most species, however, this disaffection, this first rift in the happy relations at home, is kept under control. Within a day all is well. They have accepted the position and are their normal good-natured, busy selves again. But in the home of *B. lapidarius,* that beautiful bee of jet black and brilliant red, this surly discontent often ripens into something more ugly. Normally, there is nothing about this bee to lead one to expect abnormal behavior in its workers. It is an ordinary hard-working type in spite of its smart dress. Nevertheless some of its workers do not stop at grumbling or wearing sour looks; they lose control of themselves and begin to tear open the cells and destroy the eggs. The queen rushes from cell to cell, thrusting them aside, even fighting with them, and hurriedly repairing the rents and damage. The other daughters, though they do not actively join the mutiny, give no help to their mother but stand sullenly aside. It is touch and go. What damage is done depends on how many take part in the disgraceful scene and whether the mother is able to check the out-burst. But even if things do not get completely out of hand, several eggs will be destroyed. Luckily, at the bottom, it is only an exhi-bition of female hysterics. In five or six hours it will be over and the malcontents will be back at work, probably a little ashamed of themselves.

I have singled out *B. lapidarius* as being the species prone to this disorder, but *B. terrestris,* the cutter of holes in flowers, does not altogether escape stigma, though with *terrestris* this family crisis does not usually get quite so far out of the mother's control.

And so, before long, the last of the fully-sexed daughters says goodby. Peace and quiet descend on the old home. It is a grim peace really, a rather deadly quiet after all that bustle and activity. The mother lays no more eggs. She is old. Her coat is threadbare and she feels the chill the autumn evenings bring. Her daughters barely number more than that first little batch she raised with so much care and pride. They are the remnants, the youngest and strongest—young and strong, alas, no more. Mostly they stay at home—what is there left to work for? And the flowers now are scarce; there is no incentive to hunt them out. The cold and damp increase. The mother huddles in a corner and will not leave it. A green mildew creeps over the empty cells. A wetter and more bitter night than usual does its kindly work: the mother in her corner will feel the cold no more. Her daughters crouch on the moldy comb waiting for the same release. Somewhere, in some nook or cranny, some burrow in a bank, in moss, in thatch, even in rubbish heaps, rest the princesses. They have had their spell of glamour, their brief dalliance. That is over now. A torpor has come over them, something that looks like death and will last nine months. In those inert bodies, the frequent prey of rats and mice and creeping things, lies the sole hope for the future for bumblebees. The queen is dead; long live the queen!

With *B. pratorum,* that very early bee, she of the dull-gold bands and reddish-orange tail, the death scene is less grim. Since this species begins its nest in March or April, the work is finished by June or July. With all other insects it is the time of greatest activity, of the peak of plenty, but the *pratorum* mother and her daughters just sit on the combs and relax. It must be fairly pleasant, for the nest is warm and the honey vats are full, and when, one day, the inevitable comes to pass and a daughter announces that the honey is finished, no one takes any steps in the matter, though honey is there for the taking, and in quantity, almost at their door. It may be that they are worn out or it may be deliberate suicide, but they seem to suffer no discomfort from the lack of food. A drowsiness comes over the mother, just as it did about a year ago; but the bright day in early March that wakened her before will not do so again. And to the workers too, the end comes just as peacefully.

It would have been confusing to interrupt this short sketch of the bumblebee's life in the middle, so I'm afraid we must go back now, back to the queen's nest shortly after she has raised her first bevy of children, and study a mysterious visitor who may arrive just about this time. If she does come she enters quietly, and the family, like the yokels they are, stand and gape at her. She is a handsome bee, fully as large as the mother, but smarter; her coat is not patchy and worn, and her figure is better (as mother's used to be before she spread out so much). Her colors are very similar to the queen's, but unstained by work. Had there been a door she would certainly have knocked, since she is diffident and has nice manners. Perhaps it is on account of these nice manners that she is suffered to remain, for strange bumblebees are not as a rule allowed in other nests.

Not only does she remain, but soon she gathers quite a court of admiring young females. Evidently she has charm as well as a smart appearance; they fuss around, giving her every attention. However, not all of them have fallen for her charm; one or two workers are not fascinated, and the queen is plainly ill at ease and goes about her work in a worried, agitated way as if she were controlling herself with difficulty. The newcomer, however, is tactful and keeps out of the queen's way. And really there seems no reason why any of them should dislike her; she conducts herself irreproachably, remaining quiet and unassuming and putting on no airs. In fact, she behaves just as if she were one of them—except that she does no work.

And so it goes on; very harmless on the surface, but there is a different spirit in the home—a strained, unhealthy atmosphere. There are probably not a few quarrels as well, and the mother grows more and more fretful and her hostility to the stranger becomes more pronounced. But the latter takes no offense; indeed, the ruder the mother becomes the more polite is she. All the same one would think that if she had really good taste she would go away, so obvious is the trouble she is causing—but the last thing she intends to do is to go away. And do not the bulk of the workers adore her?

Then, gradually, the attitude of the visitor begins to change. She becomes less self-effacing and makes no further attempts to win over the non-admirers, while she ignores the mother. Never actually rude, she ceases to try to get out of her way; it is the mother who has to get out of the way now, not because she is made to, but because, in her dislike, she cannot endure the propinquity of the charmer. Her few loyal daughters back her up, but the infidelity of the others must have cruelly wounded her. And it is not long before the faithful ones begin to suffer for their faithfulness. They are chivied about and have to watch their step. The newcomer eyes them malevolently and at times pursues them over the combs.

The mother is a very different creature now from the busy matron who once so cheerfully ruled a contented home. She sulks in odd corners, nursing hatred. A climax is clearly at hand, and it is a climax for which the intruder has been scheming ever since she stepped so quietly into the nest. It is not yet known who makes the attack, the mother or the murderess. Probably it is the mother, and probably when the time is ripe, she is deliberately incited to take this fatal step. Because fatal it is; as well might a naked man with a dagger attack a knight in armor and carrying a sword, as the mother attack this distant relative of hers. For the uninvited guest, in spite of her appearance, is not a true bumble. Her name is *Psithyrus* (which comes from the Greek and means "whispering," referring to her voice, which is softer than that of the bumblebee), and she has a skin like a rhinoceros in more senses than one. It is so thick and hard that the sting of the bumblebee cannot penetrate it, and when one adds that this formidable female has a strong thick sting which makes the bumblebee's look like a toy, it will be appreciated that no bumblebee can have any chance against her.

So the queen attacks and is slain, and *Psithyrus,* the usurper, reigns in her stead. At the murder of their mother the loyal daughters hurl themselves upon the killer and are slain also. They are joined by one or two of the others as the scales—too late—fall from their eyes.

The deluded workers have no alternative now; they have sold themselves into slavery and must work for an alien mistress. The eggs laid by the dead queen hatch out and are reared, and before long a large band of workers, most of whom knew no other ruler, are bringing in supplies and attending to the household duties. The usurper needed these, for she herself cannot breed workers, only sons and sexed, indolent daughters to carry on her evil strain. So that is the end of the bumblebee family. They leave no successors; only good-looking alien princesses to destroy in their turn other honest families.

To be sucessful in almost any field, insects and men must act at the right time. The general who strikes too late or too soon loses the battle. So does *Psithyrus*. If she times her arrival in the bumblebee nest too early, or precipitates too soon an attack from the queen, there will be insufficient coming workers in the nurseries to look after her and her future family. If late—well, on the face of it, there would seem to be nothing wrong with that; the nest will be populous, full of workers, her potential slaves. True, the mother by now may be rearing her own sons and perfect daughters, but a murderess can easily cope with them, and those that escape her attentions will be dealt with by her own brood. No, the snag is of a different kind and lies in a breakdown of the famous charm. It is reasonable, I think: a lady can exercise her fascination much more easily on a few than on a crowd; anyway, it takes longer to get it going with a crowd. Moreover (though don't tell her this), those first few daughters of the mother were not quite so healthy as they might have been. They were under-nourished and some of them had not received enough warmth in the hatching, and sickly females are often neurotic and apt to be carried away emotionally. Be that as it may, the schemer entering a more crowded home meets daughters who sum her up with the same discernment as their

mother. It must shock *Psithyrus* to find that instead of being irresistible she is treated as someone not at all nice to know. Her
intended hosts run about in a kind of dismay almost (what an in-
sult!) as if an enemy had appeared.

If she now tries, as I presume she does, to turn on the fascination
she has to drop that line very quickly, for soon she is fully engaged
in a fight in which practically all except the distracted mother take
part. I have said that *Psithyrus's* body is covered with armor impenetrable to the sting of the bumblebee—and so it is, but even she
has her Achilles' heel, situated, inappropriately, in the neck. That
short white thread is a minute target; the chances of its being
touched normally are negligible, but when a hundred or so are
fighting her, thrusting their stings at her blindly in every part, one,
sooner or later, will reach the vital mark. When this happens, the
would-be usurper falls and the panting workers withdraw. Victory
has been won—but at a price; fifteen or twenty, almost a quarter
of their number, lie dead or dying.

It is not often that the clever *Psithyrus* makes this mistake, but
she does sometimes, and after these too late visits her dead body and
the dead bodies of the workers can be seen in the nest. For observation purposes a captured *Psithyrus* can be introduced to an
advanced family and her reception noted.

You or I, seeing a *Psithyrus* at the flowers, would never think
that she was anything other than an ordinary bumblebee. She will
have a softer hum, and since she does no work she will have a better-
groomed appearance, but neither is likely to be noticed. The tell-
tale mark, the brand, so to speak, of Cain, lies in her hind legs,
which have no pollen baskets—also the result of never doing any
work, and also unlikely to be noticed. The very observant may
mark her lackadaisical manner; the busy bustle of the bumblebee is

absent; when a *Psithyrus* is out she is not getting the rations for a multitudinous family, but stuffing her own belly, which is probably full enough already. She sniffs a lot of flowers until she finds one whose flavor titillates her appetite, and finally, when really full, she sleeps the sleep of repletion. So when you see a "bumblebee" sleeping on a flower, you will have cause for suspicion.

What are these sinister creatures and how did they originate? There are several species, and a peculiar thing about them is that each species confines its operations to only one species of bumblebee. Never, for instance, will *P. rupestris* victimize any other bees than *B. lapidarius; P. vestalis* than *B. terrestris;* or *P. quadricolor* than *B. pratorum.* So since we have seventeen species of bumblebees, it follows that eleven of these are immune from any *Psithyrus* visits. Another significant fact is that the victimizers always wear coloring similar to the victims. Were the intruders once bumblebees, daughters that took a wrong turn and became adventuresses? The possible beginning of such a back-sliding can be seen in certain bumblebees today. Queens of *B. lapidarius* and (need I say!) *B. terrestris* sometimes enter the nest of a sister queen in just such a way as *Psithyrus.* If they can establish themselves, they dodge all the initial drudgery of a normal queen. They, too, efface themselves at the beginning and take care to be as unobtrusive as possible, and with them too the showdown has to come, and the two queens unite in deadly struggle. This time justice prevails, for it is nearly always the rightful mother who wins.

Psithyrus may have originated from such types. Females with strong objections to work are the most likely to go astray. Strains developed that grew more and more work-shy and more and more cunning in evolving methods of evading work. Such strains, breaking away from the routine of the parent stock, might in time acquire different characteristics. Selection, too, would play its part. Murder being essential for success in their new life, only the victors of fights would live to perpetuate their kind. This would eliminate those without a hard protective covering and gradually result in the armor they wear today. For the same reason their sting would become more formidable. Yes, I think we may assume that the *Psithyrus* clan are descendants of erring bumblebee daughters that took to bad ways in the dim past. As for the charm—well, ladies who live by their wits must have it, or revert to honest work.

A lot of knowledgeable people do not feel too happy about bumblebees. I have spoken to farm laborers of the old type, and they agree that nests of these bees are becoming much scarcer. About the peach blossom and early flowers there can no longer be found the same number of queens that used to appear in former years. Preyed on by numerous enemies and frequently destroyed by rains and floods, they have always led a precarious existence; now more extensive cultivation and the spraying of road verges and other places with weed killers is making this existence more precarious still. This, no doubt, cannot be helped, but the spraying of fruit trees at blossoming time can. It destroys large numbers, most of whom are queens, and the destruction of one queen at this time means the destruction of several hundred workers and the loss of fifty or so of next year's potential mothers.

This is a tragedy, not only to the bees but to ourselves. Bumblebees are invaluable to us; after the honeybees they are our chief pollinators and, at times, our only pollinators. The two together are a grand combination. The honeybees, to a certain extent, we can protect, but the bumblebees must work out their own salvation. In an old-time song a negro, treed by a bear, lifts up his voice for succor from on high and, lest his prayer may not be granted, concludes with the request, "Oh, Lord, if you can't help me for goodness sake don't you help that bear." We cannot *help* the bumblebee, but we can refrain in many ways from allying ourselves with its enemies.

CHAPTER EIGHT

THE SICK BEE

Insects, if we are to judge by the bee, suffer from a variety of diseases, many of them akin to those terrible plagues that used to sweep through our cities in the Middle Ages. They indulge also in minor complaints and are particularly prone to stomach troubles such as dysentery and diarrhea. Like ourselves, they have their infantile diseases, some of them much more virulent than any we know. One in particular is a veritable "Black Death": the pearly, glistening maggots turn black and die at greater speed than the undertakers can dispose of the bodies. It is known by the unenticing name of foul brood. Dead maggots hardly excite us to pity, yet the bees suffer great distress when their grubs get ill and die. There is a peculiar complaint—an adult one—which causes those affected to lose their hair and become definitely mental. The bees—who have no time or room or use for lunatic asylums—are kept busy fighting and throwing out their bald and mentally afflicted sisters. Naturally, it is only the more obvious diseases that the beekeeper can spot. I have no doubt insects suffer from minor complaints just as we do. The fly you swat may have had a headache that morning due to overindulgence on fermenting fruit, or a cold through being out too late.

Be that as it may, a description of bee diseases is really only suited for a textbook on beekeeping. Furthermore, brother beekeepers will be annoyed with me for mentioning diseases at all. They have an idea that the public should not be allowed to know that bees have diseases. I suppose they think it will put people off buying honey for fear of catching some mysterious bee complaint, or having their babies turn black. As well might the public refuse to buy bread because the baker's infants were not immune from mumps. This is not a good illustration because there is a certain amount of logic in it. One *can* catch human diseases in many devious ways; under no circumstances can one catch a bee disease, unless one is a bee oneself, or at any rate an insect.

Still, as I say, only the man who keeps bees is likely to be interested in their complaints, but there was one particular complaint that was different, and I am going to deal with it now. It was a disease that had far-reaching consequences and many unexpected turns. Moreover, strictly speaking, it was not a disease at all; it was an attack by one insect on another.

In 1904 beekeeping in Britain was booming. A new type of hive and various recently invented gadgets had enabled one hundred or two hundred pounds of honey to be gathered where but twenty or thirty were obtained before. The market, in short, was "active," and the sale of bees alone an extremely profitable side line. In fact, at the rate things were going, a slump was bound to come. Bees, like cattle, must have pasturage. There is always a saturation point. Knowledgeable beekeepers foresaw and dreaded this slump. The slump came, but not in the way they expected. It came from lack of bees and not from too many.

I remember being impressed in my youth by the story of the prophet in Roman times who kept calling out in the arena. At that time mass executions of Christians were extremely popular. So much so that even the vast amphitheater of Rome could by no means hold all the spectators. The result was confusion and a lot of treading underfoot of spectators and a lot of protests at the bad management to the Emperor. Loud above these protests, every day, rose the voice of the prophet spectator: "In but a few days more there will be room and to spare." There was an ominous edge to these words which came true when Rome was smitten with a fell disease and the amphitheater became half empty. Incidentally, the prophet was executed. No such voice called out to beekeepers at the beginning of this century, but if it had it would have been equally apt. A fell disease was on its way which threatened no less than the complete extermination of the honeybee.

It was in Newport, Isle of Wight, in this same year of 1904, that a beekeeper, talking with fellow beekeepers, mentioned the peculiar behavior of two of his hives. No one took much notice; all were too busy talking about the peculiar behavior of some of *their* bees, for bees are always doing peculiar things according to their "owners," and the "owners"—to hazard a guess—do even more peculiar things according to the bees. However, a few weeks later, one of the beekeepers of the party asked the first to state exactly what it was his bees had been doing. Boiled down, the first beekeeper's story was that his bees had got very fat and were holding processions— large processions that marched every day steadily and orderly out of the hive and continued their progress like a moving carpet along the grass. They seemed full of the joy of life, for as they marched they waved their wings exuberantly and, like schoolboys, climbed up all the tall grasses they encountered. This would go on for some time, and then a more sober attitude would come upon them. They would split up into small groups that would crowd together on any raised place such as a stone or clod of earth, and there stand heads bowed down as if in prayer. The symptoms really were not unlike those of some religious revivals, except that there was no singing. The second beekeeper said that it was strange, but three of *his* hives were doing the same thing, and he was worried about it because none of these processions ever returned.

Whatever was the cause of this behavior among the bees it caught on. The next year almost every apiary in the Isle of Wight had one or more of these peculiar hives, and the year after that there was hardly a stock which was not, so to speak, walking out on itself. Obviously something was amiss. The Board of Agriculture, badgered into it, sent an inspector down, and later, on his recommendation, a scientist. In 1908 the scientist was able to report that all the original stocks of the Isle of Wight were exterminated, and that stocks imported since were in the process of being exterminated too.

The Isle of Wight is, normally, a good place for bees, and large returns are obtained. But few people now envied the beekeepers who lived there. That was all it amounted to; the rest of Britain said it was just too bad for the Isle of Wight. We should feel the same if we heard there was some peculiar complaint among the sheep on some island, mildly sorry for the island. That attitude, however, changed very abruptly in 1909, when bees affected with what was now known as the Isle of Wight disease were found in Hampshire, and almost immediately afterwards in Sussex. After that the plague spread like a prairie fire through the whole of Britain.

The Board of Agriculture, thoroughly roused at last, appointed a board of scientists from Cambridge to investigate. Their investigations took a long time, but in 1912 they issued their report. There is a well known bee disease called nosema. The scientists said that they had found nosema spoors in many of the affected bees, and that nosema was the trouble. They did not put it quite in that way, for it was an exceedingly long report, but that was what it amounted to: they had found the disease, and it was nosema. The only other thing of real interest in their report was a suggestion that the disease should now be called microsporidiosis. One can well understand how a word so easy to pronounce as nosema must have grated on some scientific ears. Having come and seen and conquered, they then went back to their normal activities in Cambridge.

Quite what good they thought they had done it is difficult to understand. No cure had been suggested or even looked for, and bees were still dying in vast quantities all over the country. Moreover, a lot of people were puzzled. When learned men state a fact in a voluminous report one feels it impertinent to doubt their word. The Isle of Wight disease was nosema. They had said so. And nosema spoors had been found in the stomachs of infected bees. Quite. But why was it that other bees obviously and definitely in the last stages of Isle of Wight disease were free from any spoors of nosema? That had not seemed to bother the Board of Scientists nor, strangely enough, the Board of Agriculture. But, as I say, it bothered some people. Among these were two Scotsmen, and the more they thought about it the more dissatisfied they were, until, in the end, they decided to investigate on their own account. But money was needed. Investigations of this nature may take years,

and scientists must live. It was no use applying to the public. The
public is not interested in bee diseases. All the public wants is honey,
and it thinks that a creature like a bee has no right to have a disease in any case. Luckily, in the end, a Mr. Wood of Glassel stepped
into the breach and provided the bulk of the funds. Drs. Rennie
and Anderson of Aberdeen University then began their investigations.

They started full of hope, but at the end of five years they had
accomplished exactly nothing. The disease eluded them. No germ
or spoor of any kind could be found except, occasionally,
nosema. And they still held to their first opinion that, if nosema
was the cause, its spoors would be found in every bee examined.
It says much for their dour northern persistence that they went on.
But they did. At least Dr. Rennie did. Dr. Anderson may have done
so also, but his name drops out of the later reports. Then came one
of those chance discoveries that so often (more particularly in
chemistry) have proved a turning point. A Miss Harvey, one of Dr.
Rennie's assistants, while examining a dissected bee, found a mite
in the upper portion of a breathing tube. No one thought anything
of it, and she less than anyone else. In fact she did not mention
it for some time. Mites are often found on bees, though none so far
had been found *in* them. In any case the symptoms of Isle of Wight
disease were swollen abdomen and dysentery. Naturally the germ
was looked for in the stomach and intestines. The inability to fly
was put down to general weakness, for flying to the bee—with its
small wings—is a feat of tremendous energy, demanding eight
times the exertion of walking. By now Dr. Rennie feared that the
germ was one of those baffling filter-passing organisms such as defy
detection in several human complaints. He was interested when he
heard about the mite, though he too regarded it as of no importance. In his spare time he looked for others. He found them, and

gradually the fact emerged that every bee infected with Isle of Wight disease had mites in its top breathing tubes. At long last they were on the scent.

Probably the truth dawned suddenly on Dr. Rennie. These mites, and not a germ, were responsible for the disease. The mites invade the two big tubes in the upper portion of the thorax. These two tubes supply oxygen direct to the wing muscles. If the tubes are choked with mites, the wing muscles can get no oxygen. Therefore the bee cannot fly. As for the swollen abdomen and the dysentery, that also became clear. The cleanly habits of the bee were responsible. A creature that cannot tolerate a speck of dust naturally does not allow the voiding of feces inside the hive. Even in the cold of winter a bee must wait till it can go outside. As a result of this age-old prohibition the bee has become practically incapable of voiding its feces except on the wing. When it cannot fly, therefore, its waste matter accumulates and dysentery ensues. It is at this stage that the bee walks out of the hive to die. The first few bees are unnoticed, but once it has got a footing the mite breeds rapidly. Pregnant females in numbers emigrate from the first bee infected and invade other bees, from each of which in due course emerge other bands of pregnant female invaders. The whole hive soon becomes infected with varying stages of the disease, so that every day large numbers of those about to die march out.

The problem was solved, though it had taken twenty years to solve it. In honor of the man who supplied the funds the mite was called *Acarapus woodi*. The indefatigable Dr. Rennie now set out to find out all about its life history. It was a difficult task, but he succeeded. We will not go into it here, but we may as well mention one or two points about this hitherto unknown creature. It is a mite a quarter of the size of the common cheese mite, and therefore invisible to the naked eye. Though not an insect at all in the scientific

sense, it is very like one, having a breathing apparatus similar to
that of the bee, laying eggs which turn first into nymphs and then
into mites. It has six legs and a pointed snout, with which it can
puncture the lining of the tube wall and suck the bee's blood. In
place of a tail the female has one long hair and the male three. The
males are the stay-at-homes; only the females migrate and found
fresh colonies in other bees.

The problem was solved, certainly, and scientists everywhere
(except, presumably, the Cambridge ones) felt pleased about
it. Beekeepers were not quite so pleased. One supposes that they
were interested in the discovery, but one could hardly expect
them to be exactly jubilant over the fact that their bees were
dying from mites in the breathing tubes instead of germs in the
stomach. The sad fact was that beekeepers (and bees) were no
better off than they had been before. What *they* wanted was a
cure; something which would kill the mites in the lungs without
killing the bee, not an easy thing to find. To get a picture of the
difficulty and the comparative size of the mites, you must imagine
a man with his bronchial tubes full of cockroaches. Something
has to be found that that man can breathe in to kill those cockroaches. There are many agents which would do this—but they
would also kill the man. An added difficulty was that the *eggs*
of the mite must be destroyed also at one and the same time. Dr.
Rennie now set himself the task of discovering a cure. We have
seen that he was no defeatist; that he went on and on. He did
so now, yet in the end he was forced to announce definitely that
there was no cure for Isle of Wight disease—or acarine, as it
had come to be called. Science, in short, threw up the sponge.

Then came another surprise. Where scientists with their knowledge, apparatus, and paid time failed, a stationmaster succeeded.
A Mr. R. W. Frow, stationmaster of Wickenby, Lincs., announced

in a bee journal in 1927 that he thought he had discovered a
cure. He gave the recipe, too, and a remarkable mixture it was,
including petrol, nitrobenzine, and saffrol oil. The idea was to
apply it to the hive on a pad so that the fumes permeated the
whole hive. The bees (he said) would breathe it and (perforce)
the mites, and the mites would be killed and the bees would not.
Not only this, he claimed that it would kill the eggs of the mites
as well. A number of beekeepers tried it immediately, including
several of that class who think that if a stipulated dose of medicine
is doubled it will do twice as much good. I had a friend like this
who had a dog with worms. He bought a strong medicine, starved
the dog for twice as long as the instructions directed, and gave
four times the required dose. He was fed up with those worms
and wanted to kill them properly. Whether he did or not he
certainly killed the dog. He then wrote an abusive letter to the
proprietors, and warned all and sundry against the medicine in
question. That is exactly what happened with Mr. Frow's remedy.
Certain people made doubly sure of killing the mites, and then
wrote to bee journals saying that they found all their bees dead
the next morning. Those who gave the correct dose were not in
a position to give an opinion. Even on bees dissected and examined
it is a month before any result can be seen. It may be four months
before the ordinary beekeeper can be sure his stock has been saved.
Once these overdosers gave their verdict the usual crowds who
decry anything new arose and in various papers poured scorn
on Mr. Frow. To make matters worse, Mr. Frow's first suggested
dosage *was* a little on the strong side, and some stocks died. The
one and only cure for acarine disease went within an ace of being
stillborn. Then the third of the trio who saved our bees arose. This
time it was an army man (a strange trinity: a scientist, a stationmaster, and a soldier). The task of Colonel Holsworth of Devon-

shire was less spectacular, but it came at a crucial time. Instead
of joining the pack now in full cry running down Mr. Frow, this
private beekeeper conducted a series of the most painstaking and
exact experiments. He sacrificed his stocks wholesale, but in two
years he found that the cure *was* a cure, and he ascertained the
correct dosage—and it is a hair's-breadth matter when both the
mite and its host are breathing in a poison: fortunately the mite
is a little more susceptible. And here it must be added that Mr.
Frow—who could have made a fortune—gave his discovery to the
world free, gratis, and for nothing. Already acarine was appearing on the Continent. It would have spread. But for him the
honeybee—in the Old World at any rate—might have become
extinct.

And so, in twenty-five years, the mysterious bee disease that
first appeared in the Isle of Wight was conquered. The question
may be asked, if the disease spread so fast and was so devastating,
how was it that any British bees survived? The answer is simple:
none did. The bee that worked in British gardens fifty or so years
ago is with us no longer. It is a pity, for it was a splendid bee and
knew the climate as no bee knows it now. For while the disease
was striking down British bees, beekeepers were busy importing
other races. French, German, Russian, Italian, and other stocks
of bees were shipped to this country in large quantities. This was
not only to keep up the supply; the idea was prevalent that since
it was *British* bees that had been attacked, the bees of other countries might be immune. And for every new race imported, the cry
went up that a "resistant" breed had arrived. Of course, it was
only a matter of time. *A. woodi* cares nothing for the nationality
of its host; it invaded one and all, and when the jubilations were
reaching their loudest the dreaded symptoms appeared and the
processions started. The finding of a "resistant" bee was all the

rage then. Actually the bee has as much hope of being resistant
to the mite as a man in a vermin-ridden doss house has of being
resistant to fleas. Lastly came Dutch bees, and this time the cry
went up so loudly that a resistant race had arrived that the Govern-
ment took a hand and brought them over by the shipload. They
did seem immune, too. While other stocks were dying they remained unaffected, and great was the rejoicing. But at last they,
too, began to "crawl." Their seeming resistance was due to the
fact that they bred so fast and swarmed so much that they almost
kept pace with the ravages of the mite. Not quite. The end, though
protracted, was equally sure. So now we have a mixture of all
these races. British blood is there, but nowhere is it pure, and
nowhere can be found the old, original, hardy British black bee.

The origin of this wretched mite is still unknown. Actually it
was only due to a slight lapse on Nature's part that it was able to
attack bees at all. Bees (and all insects) have a far more efficient
breathing apparatus than any animal. Animals breathe through
one small tube which is also used, in part, for the conveyance of
liquids and solids—a very poor and rather dangerous arrangement. Insects breathe through a whole crowd of tubes relegated
to that purpose only and which do not interfere with the mouth
and throat. This is why insects have such amazing vitality. These
tubes are mostly in the thorax, and are very small except for the
two main tubes called the tracheæ, which have been mentioned
before as supplying oxygen to the wing muscles. The openings to
these two tubes (called the spiracles) are also comparatively large,
and therefore are a possible source of invasion by germs or very
small mites. To guard against this they are provided with a
'shutter,' which operates much in the same way that we sneeze.
Now with bees this shutter does not operate properly until they
are five days old. After that it is perfect. I cannot say why this

should be, but it is so. And those five days are the mite's opportunity. A bee over five days old is immune from acarine—provided it has not already contracted it. But once *A. woodi* is in a hive those five days are quite sufficient for its purposes. To go back, where did *A. woodi* come from? It was in Britain somewhere in 1904, and it had already developed its life cycle in the interior of some insect—but not, until then, in the bee. One thing is known, it cannot "crawl about." It can only invade another bee by direct contact, shoulder to shoulder, so to speak. Dr. Rennie examined insects of every sort: wasps, bumblebees, flies, ants, and others. In the tracheæ of none did he find any trace of invaders. Still, this proves nothing. I think we may assume that somewhere in the Isle of Wight, some time round about 1904, a bee rubbed shoulders with some insect that was the normal host of *A. woodi*. Probably it was in a spider's web. At any rate it was a chance encounter that might not have happened again for another thousand years, if at all. By a mere fluke like this was the British bee exterminated.

I well remember the time I found acarine among my own bees. It was my first full season of beekeeping. I had studied bee literature intensively and bought a microscope—quite a good one, though all I had done with it so far was to study bees' legs and things like that, for which it was eminently unsuited, the lowest-powered lens being much too strong. Later that year I attended a course of beekeeping at an Agricultural College, and with colossal cheek joined the advanced class instead of keeping to my proper sphere with the beginners under a lady mistress. In this class I learned to dissect bees for acarine and nosema and to stain slides and spoors. It was three days after this course that, looking at my brown hive, I saw a bee trying to get home by crawling up a blade of grass and signaling wildly with its wings. It was a

The Invalids

very fat bee, and one wing stuck out at right angles. Then another
bee walked out of the hive in a meditative sort of way and fell
off the alighting board. This one also climbed a blade of grass,
tried to take off, and fell to earth again. On a stone nearby were
three bees, heads close together, a motionless group. These were
the external symptoms of acarine—the old, notorious Isle of Wight
disease. Yet with five bees only it was nothing to go by. A slight
attack of ordinary dysentery due to unsuitable food or a damp
hive, poisoning through contact with sprayed plants, wounds,
old age: all these would produce similar symptoms. The only

sure external sign of acarine is what beekeepers call "mass crawl-ing"; in other words, large-scale emigration on foot—though the one wing stuck out at right angles is always a little bit suspicious. I put one of the motionless three into a matchbox. Beyond moving its legs very slightly it did not seem to mind where it went. I put it into a small glass tube containing bits of blotting paper soaked with nitrate of ether. This is supposed to be a painless anæsthetic and killing agent, but its effect on that bee was astounding. I expected to see the already barely animate form go limp and life-less; instead, the bee reacted as one restored to the full vigor of life. It made that small tube literally hum, and it dashed itself about so rapidly I could hardly see it. After this I no longer used this alleged painless method of killing the bees I dissected; I simply jerked their heads off with a pair of forceps. The bee died eventually, and I pinned it to a cork and began to dissect. I took off the head, the two front legs, and the "crown," and exposed the chest; then looked for the two main lung branches, or tracheæ, with a pocket lens. I saw them, tiny filaments. At the college I had got them out and on to the glass slide into a drop of water or methylated spirit after about six attempts. It took me even longer on this occasion—and then I failed! With infinite care I *seemed* to get that filament on to the glass slide, but all I saw through the lens was a string of muscle. Under the microscope the trachea is quite unmistakable. It is a great tube reinforced with spiral wire like a powerful delivery hose. I carved up the three on the stone to no purpose except to give them a quicker release, did the same to the bee on the blade of grass, and then hunted for the last one. It took me a long time to find her, for she was on the move, but I spotted her at last. And this time I got the trachea out intact. I pressed the top slide over it, adjusted the light, and focused it. I moved the slide slowly, and the great corrugated pipe came into

view (the way the microscope transforms the insignificant into
something massive, detailed and wonderfully finished, seems to
me quite as remarkable as the fairy wand of stories). I made the
slide travel slowly, and the pipe passed along the lens like some
endless structure. And then I stopped. In the center was a moving
creature: a thing like a tortoise, that waved six legs in the air.
The legs were thick and puffy, and each was armed with three
long tapering claws. It had a pointed snout like a drawn out pig's,
and a long hair in place of a tail. Here in the flesh, alive and kick-
ing, was *A. woodi*. The dread infliction of the Isle of Wight had
come upon *my* hives!

I do not know whether sorrow at having the disease or jubilation
at having myself, alone and unaided, dissected and found the
living cause possessed me at that moment. On second thoughts,
I do. It was jubilation. I know I fetched my wife to come and see
it, which she did, and when she had inspected the apparent
nothingness that was on the slide beneath the lens, kindly said that
she thought it was very clever of me to have found it, with which
I agreed. This was a single female invader, and could not itself
have caused the distress and illness of the bee. There are two of
these pipes. I had only the one. Doubtless the other was choked
with mites.

You may ask, why all this fuss? This Mr. Frow has apparently
invented a cure—go and apply it. It is not quite so simple as that.
Firstly, if the mite is in one hive it is almost certainly in the others
as well; all must be treated. Secondly, the highly volatile fumes
of Frow's mixture have (as can be imagined) a stupefying effect
on the bees, and stupefied bees are not much use at guarding their
stores. The prowling robber bee that is always with us detects
the strange scent, and being a curious as well as an evil customer,
investigates. To her delighted surprise the guards seem to be in

a sort of trance. She winks to herself as she slips in, gets a load
of honey, and goes on her way rejoicing. An hour later the treated
hive will be hardly visible for the enveloping crowd of shrieking,
looting bees, and next day there will be nothing much of it left
to treat. So the bees must be shut in for at least a fortnight, and
one can only do this in winter. Thirdly, a "cure" for bees is not
what *we* would consider a cure as applied to ourselves. It is as if
a man went to a doctor and said he felt ill, and the doctor examined him and said, Yes, he had such and such a disease, but
he must not worry because there was a splendid cure for it now.
He would die in agony but his children yet unborn would be
saved if he took this bottle of medicine and used it as directed.
This is all the consolation the bee gets. She must bring up young
quickly to take her place, for she herself is doomed. To bring up
and feed young in sufficient numbers to carry on is exacting work.
A badly disease-ridden hive could not do it. So it is only those
who are not too far gone by winter that can be saved.

The brown hive was not one of these. Still, I got a lot of fun
out of it, and I saved the other. At the first opportunity I shut
them all up and began the treatment. The treatment over, I
began dissecting and examining bees. I could not say how many
bees I dissected daily, but I became marvellously expert. I could
extract the trachea and have it under the microscope intact in a
jiffy. The results of the Frow treatment are not immediately
apparent. I was heartbroken to find the first lot full of quite lively
mites. In others, a few days later there were still live mites, but
there were also those with legs stuck rigidly out who showed
plainly that they had received the dread summons. Later speci-
mens presented the inspiring sight of dead mites and dead yellow
mites, the yellow ones being those that were beginning to decay.

The mite has the extraordinary capacity of laying eggs larger than itself. I do not quite know how it does this, though it was explained to me by a learned professor at the college. Under the microscope, in water, these eggs shine large and translucent like pearls. And when the mites were dead and decaying, they still did so. I got worried. Eggs do not breathe like the mites themselves. Had the fumes affected *them?* Then as the days went by they, too, began to go yellow and rotten and to break up. It was a complete rout. The invaders were destroyed. Male and female, young and eggs, root and branch, they had been annihilated.

As I say, I was too late to save the brown hive. Mass crawling developed, and finally a dose of cyanogas put an end to this stock, which had given me nothing but trouble since I got it. Their ill temper, however, was explained. Bees with acarine, the mites puncturing the tube walls with their sharp beaks, must suffer the tortures of the damned. No wonder they are irritable. Toward the end they were submissive, too broken to sting or to care what happened to them.

Killing off a stock—and I have had to kill off several for one cause or another—is not a pleasant task. Some fifty thousand intelligent beings are to be cut off. A crowded and busy city has to be massacred down to the last inhabitant. In the old days of skeps they did it with burning sulphur—a frightful agent. I use cyanogas, a powder quite awe-inspiring in its immediate effect on bees, which are tough creatures to kill as a rule. At night the hive is closed, and the resonant note of alarmed bees arises. Through a small opening a dose of the powder is pumped in. The sound ceases, as if a hand had been placed over an echoing string. There is silence; then, like rain, the patter of small bodies falling on to the wooden floor.

Samuel Butler, in his "Erewhon," tells us of a mythical land where the people had strange laws and customs. Crime was looked on as a misfortune, and criminals were tended in hospitals and sent gifts of fruit and flowers by sympathetic friends. The sick went about furtively, dreading detection, arrest, and punishment. In this latter respect Butler might well have had the honeybee in mind. With them also sickness is the unforgivable crime. The sick bee knows this well, and sooner than suffer the disgrace of detection usually leaves the hive of its own will. It must be a terrible decision to have to make: exchanging the warmth and teeming multitudes that are as its lifeblood for a lonely, lingering death. Some indeed (usually the very young) flinch from it. Their shrift is short. The other bees drive or carry them off. One dislikes bees when one sees them doing this. It seems so callous, so opposed to our own notions. We forget how necessary it is. In the teeming hive there is neither room nor time for sickbeds, and infection spreads rapidly in such tightly crowded conditions. Moreover, winter always looms ahead; the winter cluster would be fatally handicapped in its fight against cold if part of its forces were diseased.

Actually I am sure that bees are far from callous. When they are trapped (by mosquito netting, for instance, put over the hive to prevent robbing) you will see them gather in little groups, the full ones feeding the hungry ones. And there are individual friendships. They are hard to discover in such a huge population, but they exist. Once I found about ten bees that had been cut off from the others in the hive by my manipulations. I set them free, but in doing so pinned one by the leg under the end of a frame. The others were scuttling off to safety, but as the trapped one let out a shrill yell one of them turned and came back. The rest took no notice, but quickly made good their escape. The friend (for it must have been a friend) did everything she could to get her unfortunate colleague to follow her. She tried to coax her away, and later stroked and fondled and fed her. At last, my curiosity satisfied, I released the victim, and the two made off for home and safety.

I have heard it instanced as an example of the bee's callousness that it will drink from syrup or water in which a comrade is drowning. But there is a fallacy about this sort of argument. It premises that the bee's brain functions, like man's, almost exclusively in conjunction with the sense of sight. Admittedly the bee is richly endowed in this respect. She has a pair of eyes for field work, which, with their thousands of facets, enable her to see fore and aft and sideways at one and the same time. She has also three additional eyes for close work in the darkness of the hive. But these, with a few unimportant exceptions, are the only uses to which she puts her five eyes. She has other senses of communication and perception, but these may well have their failings. Bees probably are unaware of another bee immersed in liquid. Whether they would bother if a bee *was* drowning I am not prepared to say. I do know, however, that they will lavish every attention on a half-drowned bee after it emerges. They will lick it, stroke it, feed it—but not indefinitely. It is advisable for that half-drowned bee to recover fairly quickly. Otherwise those ministering to it may suddenly decide it is not worth while, and carry it away and drop it. Even when they resuscitate it, its troubles are not over. It has to get back into the hive, and the guards there may not agree with the verdict of the first-aid people. For one thing, the bee will be frightened and demoralized after its experience; it will be crouching and humble, and guards are always suspicious of such. Robbers often play that game. I will give my experiences with one bee I rescued from a jar of syrup set to trap wasps. I

dropped it on the alighting board amidst a crowd of foragers.
They turned aside immediately and licked it clean and got it
shipshape. For the sake of the syrup, you will say; but this is not
so, they will do the same for a bee soaked in water or some disagreeable fluid such as Jeyes. Then they left it, and the bee
walked slowly up towards the aperture. A couple of guards rushed
up as it drew near and halted it. They circled round it, prodded
it in various places, conferred together, and prodded it again.
Finally, they walked away. This bee seemed all right to *them*.
But that was only the beginning. The bee moved on towards the
aperture, and two more guards spotted her. The same prolonged
inspection and prodding took place, during which the first two
guards came back and joined the others in conferences and prodding. Then all four went off. The bee, the living image of humility,
again moved forward, its weary eyes on that inviting black hole
beyond which lay warmth and friendship. And now a single
officious guard (probably just promoted) ran up and with hardly
any preliminaries turned her down. It attacked her, bit at her
wing, and pulled her away from the hive. Two guards close by
joined in. They did not worry about the *pros* and *cons* of the case
at all. Like policemen the world over, they assisted their colleague
without question. The bee was dragged past the other four guards,
who said not a word and were not even interested. The wretched
creature, now a condemned criminal, was rolled to the very edge
of the alighting board. If she had been thrown over it would
have been the end of her. It was wet on the ground and muddy.
Even if she had regained the hive afterwards she would have
been in such a state that no bee, guard or otherwise, would have
tolerated her. Luckily, the guard suddenly saw what it thought
was a suspicious-looking bee, and ran off to intercept it. (Incidentally, it was a heavily laden forager, whose tired movements

deceived this officious but inexperienced guard.) The other two lost interest immediately, and walked away, wondering what all the fuss had been about, anyway. Again the bee directed her steps towards the now distant aperture, moving slowly from one inspection to another. The officious guard came back, but was not so officious now. An indignant forager had taken it down a peg. The bee reached the aperture and disappeared inside. And as she disappeared, the nearest guards decided they had been wrong after all. They ought not to have passed her. Three or four dashed inside to drag her out. But the bee, once out of view, had made a flying leap for safety. By now she was hidden amongst a crowd of workers. At any rate the guards returned without her.

Newly emerged bees are rather helpless. They need feeding and looking after till their chitin covering is set. The older bees are fond of them; not so fond of them as they are of the grubs, on which unenticing objects they lavish an almost doting affection. But they are nice and kind to young bees. For instance, a very young bee will be allowed to enter any hive, though an older bee, not of that colony, would be killed. I have had bees hatch out in an upper story of the hive shut off from the rest. I have let the bees come up from below and seen them carry these youngsters downstairs. I thought I could count on this, and on the next occasion, when a small batch of young bees had been hatched among the combs at the top of the hive, I again let the old bees come up to take them down. The hive was open at the top. The bees came up and duly seized the youngsters, but instead of taking them down they began flying off with them. For a moment I thought they were going to take them home *via* the front entrance. Instead, they took them all a field away and dropped them into the grass. I had left those young bees there a little too long before letting the bees attend to them. They had suffered from cold or hunger, or both, and in the mature consideration of the others were now classed as unfit material.

If bees have a horror of the sick, they have an even greater horror of a corpse. There is no need to go round with a bell to tell bees to bring out their dead. They will do so without delay under any circumstances. When one moves a hive from one locality to another one puts perforated zinc over the entrance, sends it by train or lorry, and places it on the new stand. By this time, what with the jolting and the imprisonment, the bees will be in a perfect uproar. You will see an apparently maddened crowd surging at the zinc barrier. That crowd will appear to be out of all control, yet first in their waiting ranks will be the undertakers, each with its corpse (fatalities in these removals are always rather heavy) lying beside it, and when the zinc is removed they, with their burdens, will come out first. This business of removing the dead at the first opportunity might almost be described as an obsession. I have seen bees that I have killed myself, by accident, taken away immediately, under my nose, by bees I had thought completely demoralized by my manipulations.

CHAPTER NINE

ENEMIES

It seems to me that the bee is the victim of one of Nature's greatest injustices. That peace-loving creature exists today only because it is armed. It carries a weapon, but (with one exception) uses it only in self-defense—or what it imagines is self-defense—and the protection of its home. Why, then, has Nature decreed that the bee must die if it uses this weapon in so justifiable and necessary a cause; and die, moreover, a prolonged and painful death? Fight and you die, says Nature to the bee: don't fight and your home and kin will be destroyed. So the bee fights and dies, and the colonies endure; but it is a hard edict. Without anything at all to worry about except the weather and its effect on their source of provender, life for the bees would be sufficiently hazardous, but

with a host of enemies arrayed against them bent on massacre or
loot, or both, their lot is not a happy one. Why did not Nature
arm those strong, bulky males? They could do the fighting—
they have nothing else to do—and when they are killed who
cares? So much the less work for the females later on. But no,
the females must do the fighting as well as the work; and they
must feed the males too, who will look on with unconcern while
their womenfolk are set upon and slaughtered.

The average uninstructed human being, noting the powerful
roar and bustle of a hive, and running for life when one or two
workers come buzzing round his head, probably feels sympathy
more for those rash enough to range themselves against the bee
than for the bee itself. Yet actually, all too often, the enemies
have it entirely their own way.

In making out a list of those opposed to the bee, it would be
advisable to have some sort of order; so we will start with the
mammals, at the head of which stands the greatest, the most formidable enemy of all—man.

Until man discovered fire he was probably only a negligible
opponent, but when he discovered, or, rather, learned to control
fire he became dangerous. He smoked all the nests he found until
the occupants were suffocated, and then took the honey and the
grubs, the two making a satisfying diet of carbohydrates and
protein. (I have never, myself, tasted bee maggots, but I am
told they are delicious. Many of my colleagues in the veld in
Africa, when stores were low and game scarce, used to eat the
maggots before the honey.) And he grew cunning in marking
out the lairs of bees, assisted in certain countries by a traitorous
little bird that led him to the nests with ulterior motives.

Man became an even more dangerous enemy when the bright
idea struck him, in the dim and distant past, of hanging up hollow

logs of wood or suitable earthenware receptacles to entice swarms
and save himself the trouble of looking for wild nests. Thus
emerged the first beekeeper, and for thousands of years, until
very recent times, his methods hardly changed. What change
there was, was for the worse. The primitive barbarian with his
hollow logs of wood did less damage to the bees than our grand-
parents with their straw skeps. The barbarian destroyed all his
stocks each year; our fathers and grandfathers destroyed only the
heavy skeps and the light ones. They argued that it was advisable
not to destroy all the stocks, but to keep a certain amount over
to make honey and new colonies the next season. In this they
were quite right, but they went a foolish way about it: they kept
only those skeps which were too light to yield a large amount of
honey, but heavy enough (with luck) to survive until the next
spring. So at night—a fitting time for so dark a deed—the heavy
skeps, those containing the strong stocks, the good stocks, the
best and keenest workers, the cold-weather foragers, and the
finest queens, were taken to a pit and suffocated over fires of
burning sulphur; sacrificed, in fact, at the altar of man's greed.
It was another version of killing the goose that laid the golden
eggs. The practice continued for hundreds of years, preserving
the feeble and lazy and destroying the strong. (Oh, for some
of those queens now that perished in the sulphur pits of long
ago!)

A race of indifferent honey gatherers was the inevitable result,
but what was worse, there emerged a race lacking in stamina,
and any stocks that showed signs of recovering were killed off.
In fact, man added another unjust edict: Nature told the bee
that if she fought she died, man told her that she also died if
she worked. Poor honey gatherers are invariably the weaker and
more feeble. Nature has her methods with such. Unlike man,

she has no use for them. So while man was still busy destroying
the fit, Nature decided to destroy the unfit, and sent diseases to
do it. Some of these diseases had been present before, but not in
the virulent form they assumed. They attacked bees with a vigor
that thoroughly startled beekeepers. They still do, and keep the
research stations very busy. And the way for all this trouble was
carefully paved by our forefathers.

The introduction of the movable-comb hive—which is the
hive in use almost everywhere now—made it possible to take the
honey without killing the bees, and also to tide them over the
hazardous winter months. So now man, instead of being an enemy
of the bee, became her friend and protector—at any rate, that
is what was said when this hive came into use. Whether the bee
has crossed man off the list of her oppressors is open to doubt.
It is even questionable whether man does not still head the list.
With skeps and hollow tree trunks, the bee (regardless of her
doom) could at least pursue her own way (which is the best
way for the bee). Her city was her own; she planned it and built
it as she wished. And the shape suited her. The modern hive
preserved her from the sulphur pit—but at a price. It was no
longer her own home: it was a great, draughty warehouse of
unnatural shape. However, the bee, in her eagerness to get on
with her affairs, accepts this weird contraption of hanging frames
and wooden dummies, and the rest, and makes the best of it. It
means more work, but by sealing up every chink with propolis
the place can be made habitable. But the modern beekeeper
will have none of this. At frequent intervals he breaks open the
seals and tries to take a hand in the internal affairs of the hive;
for he has a new craze now, he is trying to stop the swarming
instinct just as his fathers tried to encourage it. He has several
methods—and a new one comes out almost every year—but they

are all alike in seriously interfering with the bees' domestic arrangements. He interferes too, continually, while nectar is coming
in; he prises open the sealed-down roof and puts on boxes full
of empty frames. It is a challenge the bee rarely refuses (though
she does sometimes); she fills this space, and when it is full—
before, in fact—comes another on top of that, and then another
and another while summer and flowers last. And in winter the
honey is taken away by the modern beekeeper and he feeds the
bees sugar and water to last them through the winter.

Well, you will say, this is not a very serious indictment. The bees
do get their stores for winter. Unfortunately, for every conscientious beekeeper there are ten who take all the honey and then give
insufficient sugar for the winter—if they could see what went
on inside the hive, if they could see the dying sisters feeding their
queen to the end of their strength, perhaps they might spare
another pound or two of sugar syrup when they close them down.
And the good beekeepers? They, of course, give their bees enough
to tide them over the winter. But honey, and sugar and water are
two very different things. The spring crop of babies will be reared
on sugar. The beekeeper will tell you that this is quite good for
them. Some will tell you that it is even better for them than honey.
I wonder! The beekeepers do not say this when they put their
honey up for sale. It possesses marvellous and unique qualities
then. And it does possess them; and we in substituting sugar for
honey are doing, to a lesser degree, what our ill-informed grand-
parents did; undermining the stamina of a race which was surely
sufficiently undermined before—though in justice it must be said
that there *are* those who *do* winter their bees on honey, and even
those who save their best honey for this purpose.

The bee has more, a great deal more, on her charge sheet
against man, including the frequent massacre of thousands by

poisonous horticultural sprays; but enough has been said—a
full list of his crimes would grow wearisome.

The badger is accused of taking honey. Whether this poor,
maligned, persecuted animal would care to add to its other troubles
by stirring up nests of bees is open to doubt. If it does, I do not
grudge it—it needs a little sweetness in its life. About its relative
in Africa, however, the honey badger, or ratel, there is no doubt.
It is not my intention to go abroad in search of enemies of bees,
but two creatures are so outstanding in their persecution of the
bee that they must have a place.

The first is the honey bird and the second, the ratel. The ratel
is a black-and-white animal, or rather grayish-white, the white
above so sharply divided from the black below that the animal
looks as if it were wearing a cape. Like man, it employs that
spy and informer, the honey bird. Once in the early morning
when I was travelling in the veld in Portuguese East Africa, I heard
a high-pitched, throaty chattering. The sound was coming toward
me, and a moment later two black-and-white forms emerged
through the grass. They seemed to be quarrelling as they waddled
along, and did not notice me until they were almost at my feet.
They stopped then and looked up at me, but a moment later resumed their dispute. They were male and female honey badgers
and, characteristically, it was the female who was doing all the
scolding. What her husband had done I don't know, but she was
seriously annoyed.

I stood perfectly still and they paid no more attention to me. A
little later I heard a bird some distance away in a tree. It was tweet-
ing and chattering and obviously excited. I only got a glimpse of it
once, a drab little bird, but by its cries, a honey bird. It, too, seemed
annoyed. Doubtless the ratels, slow enough at the best, were too
engrossed in their domestic differences to take the expedition seri-

ously enough to satisfy their guide. However, in the end the ratels
made a slight detour round me and continued on their way, and
gradually the scolding faded into the distance. I was after ants that
morning and I wish now I had followed the marauding expedition
instead; it would have been much more interesting, and anyway
I never saw the particular ants I wanted.

The honey badger must be an even more satisfactory partner
for the honey bird than man; it is more destructive. When it gets to
a nest it tears everything out and scatters comb, grubs, and honey
all around; the honey bird can gorge to repletion. The ratel cares
nothing for stings; its thick, loose India rubber hide is impervious
to them. As a matter of fact the ratel cares nothing for anything or
any creature. A bulldog would have no hope against it. The bulldog would get its famous grip, but it would only be on thick, loose
hide, and the ratel would turn in its own skin and bite where
it wished. Poisonous snakes are merely a delicacy, and it will go so
far as to add to its menu that most dangerous of all animals to
attack—to whom even the lion gives right of way—the porcupine.

The ratel's insatiable appetite for honey is akin to the human
child's passion for ice cream. It is not its natural food, but it craves
for it. In fact, it has a childlike nature, and in captivity will spend
hours happily turning somersaults or playing with a ball. Its real
diet is flesh, and in spite of its endearing ways, no more pertinacious
or bloodthirsty little creature walks the earth today. In the slow,
inexorable pursuit of its prey it could teach even the weasel quite
a lot. It displays the same determination towards bees, and,
although not a natural climber, reaches and ravishes nests in, apparently, inaccessible positions. It treats "domestic" bees with
lighthearted destructiveness. One hive, with all its honey and grubs,
must more than satisfy even the ratel's appetite. But however many
hives there are, so many will it overturn, break open, and fling

around. After solid tree trunks and holes in precipices, these flimsy, man-made contraptions are irresistible. As I have said before, the African bee is far from docile even in the most favorable circumstances, and when a dozen or so hives have been smashed up by a honey badger the neighborhood becomes unhealthy. If the beekeeper has poultry, not one will be left alive, while the rest of his stock—and the servants—will have stampeded. And the beekeeper himself will be well advised to follow their example.

For a creature of its size, whom a single sting will kill within a minute, to join the opposition against bees would seem a most unwise proceeding; yet the mouse has done so, and has, moreover, gained quite an important position. It is only recently that it took this course; from the date, in fact, the modern hive came into use. This hive has usually a long, narrow aperture and a gently sloping alighting board, and it gave the mouse the opportunity it had been waiting for. It is not the aperture and alighting board alone that help; the whole interior of the modern hive might have been arranged for this animal's convenience. It is surprising through what a narrow opening a mouse can squeeze once it has made up its mind to do so. Finding itself inside the bees' stronghold, one would have thought that the next desire of this "wee, sleekit, cowrin', tim'rous beastie," so susceptible to stings, would have been to get out. But not a bit of it; and there are two reasons for this. The first is that the inside offers an ideal home: in the square corners and behind the dummies and other arrangements the mouse soon finds a spot where the bees cannot reach it, and in the felt and coverings over the frames (vests and old shirts and jumpers as often as not) it can make a snug, warm, entirely beeproof nest. The second is that it only tries this game in winter when the bees are quiescent.

And while the mouse is in residence, what damage does it do? On the face of it, not a great deal. It eats wax and honey and makes a sad wreckage of some of the combs, but the bees could repair all this in the spring: it litters part of the hive with excreta and with bits of chewed felt, but these could be cleared out later. Its real power as an opponent lies in a factor over which it has no control—its smell. Bees are profoundly affected by smells. The smell of ordinary carbolic will subdue them as effectively as smoke, or else infuriate them—one of the two—and they probably dislike the smell of a mouse more than anything else. For instance, if a swarm is put in an empty hive where a mouse has been, they will leave it. I do not blame them; I remember I had a mouse for about a week in a barrel of apples, and the stench from that barrel was appalling. So when, faintly at first, but growing stronger, the smell of mouse comes to them and mixes with the sweet smell of honey, an unrest comes over the winter cluster; and as the smell increases so does the unrest. Wintering is a difficult process for the bee—rest and quiet are essential. When a mouse is present they get neither; they move about and get chilled. Breeding, which should begin early in the year, is neglected, and by the spring, when new blood should be ready to take the place of old, there is none. Probably the stock will die off or be robbed out; it may even commit suicide by leaving the hive *en masse*.

I have had only one mouse in my hives. I had learned that an aperture of three-eighths of an inch was safe and I measured all my apertures before tucking down the hives for winter. On that account I never suspected a mouse when the occupants of one hive began to come out when the other hives were quiet. In the spring it was robbed out, and when I examined it I found the mouse's nest (the mouse itself had had the foresight to depart before warm weather gave vigor to the bees). Perhaps the wood of the entrance had warped. I never measured it again, so I cannot say. The extent of the depredations of the mouse can be judged by the fact that all appliance-makers now provide special entrance slides to keep them out.

It is hardly possible to draw up a list of enemies without mentioning the bear, if only to congratulate bees on his disappearance from most parts of the world. His strength, climbing abilities, and love of honey must have made him in his day one of their most dangerous antagonists; but to include the goat is, perhaps, stretching things a little too far. However, a goat was once an enemy of *my* bees, so I am going to include it, even if only to expose this particular animal.

Her name was Letty and she belonged to my wife. We had agreed before her purchase that she should be kept tethered in a part of the garden away from my orchard and apiary. To my mind, Letty had the eyes of a degraded satyr, malevolent and baleful, but my wife and child always regarded her as the sweetest thing and lavished a lot of affection on her. In return, sometimes, but not often, Letty gave them about a cupful of milk. Being feminine, Letty's one ambition was to get into that part of the garden where she was not allowed, and eat of the fruit trees whereof it was commanded that she should not eat. The orchard was planted with young dessert apple, pear, and plum trees, and one afternoon, not long after we had got her, Letty broke loose. She went, of course, straight to my orchard, and when I discovered her escape she had done all she wanted. Quite half the trees were stripped of their bark—or almost so—and with satanic instinct she had selected those trees I prized the most. Their bark must have contained some tonic element; for when I found Letty she was leaping on the hives like a chamois. With her four hoofs gathered together she would spring from the top of one hive to the top of another, making little

circles and aggressive, playful motions with her head before each
take-off. Three of the hives had been overturned and the others
had been rocked to and fro. The modern hive is a fearful thing to
upset; the combs are not static as in a skep, but hang loosely: when
the hive is overturned they smash and pile up like a telescoped
train.

With murder, deliberate and premeditated, in my heart I chased
her, and she danced away with fawnlike playfulness. I am sorry
now that I did so. I think, in a few minutes more, the wildly searching bees might have found her. I would they had!

Bees dislike sweating horses, and they are said to dislike red-
haired people. I know about the sweating horses from painful experience, but not about red-haired people. Unfortunately I have
never had a red-haired friend or I would have taken him—or her
(so dead was I to chivalry when it came between me and the
study of my bees)—to see my hives so that I could note the reaction
of the bees. But I once had a gardener whose hair was of the real
authentic flaming hue and he refused point-blank to do any work
near the hives, saying he couldn't abide being near bees—so there
may be something in it. However, it is with the enemies of bees that
I am dealing, and not the enemies of red-haired men.

Birds cannot enter hives, nor can they tear them to pieces. The
bees' larders and children are therefore safe from them. Only the
honey bird has solved what for the rest is an insuperable problem.
I think the habits of this bird are fairly well known by now, for they
have been described in many books. I could not say whether it is
a common species in Africa or not; for a normal honey bird and a
honey bird that has discovered a bees' nest are two entirely different creatures, and I regret to say I was only interested in the
latter. The former is so quiet and insignificant that, unless one were

an ornithologist, one might come across it a dozen times a day without realizing it. Always, when a honey bird has come to me with its imperious summons, I have had one or more natives with me; in fact, it has always been a native who has apprised me of the bird's arrival and come running up (almost as excited as the bird itself) with the news, "honey bird come." And a native has always done the trailing—I have merely followed the native. Whether it is necessary or not, I do not know, but the native keeps up a continual whistling and falsetto chattering. I dare say it *is* necessary; for the bird must often be unable to see its followers, and this ludicrous noise will tell it that they are on the job. Without the bird's own cries, also, it would be impossible to go on, for it is more often heard than seen. In fact the two, the follower and the bird, go more by sound than sight. With one exception the honey birds always brought me to a bees' nest. The exception was when I called the expedition off after about two miles. I had to get to a certain place that day and could no longer spare the time. Piteous remonstrances (or bad language) from the bird, and very nearly mutiny from the natives were the result.

As everybody knows, the honey bird always gets its rake-off from the native in the shape of a grubby bit of comb. This is not due to any sense of gratitude on the native's part, but to a belief that, if the bird's services go unrequited, it will not be a bees' nest that the culprit is guided to next time but a coiled-up puff-adder or a waiting lion.

The tit is the only other bird I know of that displays any particular intelligence in the campaign against the bee. Its scheme is simple, but it *is* a scheme, which is more than can be said for the rest. It generally operates in winter and its method is to hop on to the alighting board and tap near the aperture with its beak. In response to this summons a lethargic guard will, in due course, appear. As soon as she appears she will be gobbled up, and the tapping will be resumed. Guard after guard responds until the tit has had its fill. The tit is a small bird and the amount of bees it consumes at one session is not extravagant, but once started these visits will be made every day and several times a day, and the drain on bee life may prevent a proper wintering. A beekeeper really has no excuse for letting tits take his bees; a piece of netting would stop

The Tit

them. The trouble is the beekeeper never *does* think tits are taking his bees because the tits are clever enough not to do it while he is watching. I used to be at my hives early and late, and never saw a tit there. It was a fall of snow that gave the game away. There, in front of each hive, was the spoor of the tit and the bare mark where it had stood, tapping.

Perhaps I am wrong in giving the tit sole credit for this bright idea. Sparrows sometimes play the same game, and when sparrows start they are worse than tits. But I think the sparrows are merely plagiarists.

Before I started beekeeping I was (and still am) extremely fond of birds, and I had a particularly soft spot for a couple of pairs of flycatchers; I used to look out anxiously for them in May. They always nested in the same place: one in the ivy against a wall and the other behind a trellis by the house. (One year the latter built into their nest some string to which was attached a yellow label marked 1s. 11½d. It hung down exactly like a neatly tied-on price ticket, but the nest was not for sale.) Then I got my bees, and later, when I was watching their hive, I saw one of my friends sitting demurely on a branch. Every now and then it would fly off for a short distance, flutter in the air, and come back to its branch, and I soon saw, to my horror, that what it was doing was nabbing my bees. At that period each bee of my one hive was like a drop of blood to me, and a heavy strain was put on my affection for the flycatchers. Sometimes one would operate and sometimes more, and the flycatcher, in spite of its small size and demure, soulful expression, is a tremendous eater. In later seasons when I had more hives I did not mind—or at least, only winced slightly—when I saw a bee taken, and my old affection for the flycatchers returned.

The swallow is too quick in its movements for anyone to be able to make a sworn accusation against it. Watching them, I have had strong suspicions. But they are said to prefer smaller insects, so we will give the swallow the benefit of the doubt; though it is well known that bees sometimes chase them away. There is no doubt, however, about the butcherbird; a stocktaking of its larders reveals that it prefers bee to almost any other meat.

The domestic fowl is more sinned against than sinning. Bees and fowls seem to get on very well together until some major disaster happens to the bees—then they usually blame the fowls for it. Ducks are in a different category, and individuals become very "fond of" bees.

An amphibian creeps in here. The toad may take up permanent quarters near or under a hive, and in its quiet, unobtrusive way take a heavy toll. It systematically gathers in a large portion of those bees that, heavy with honey, fall short of the alighting board at their first landing effort. It may live and feast there for months without even the most observant beekeeper spotting it.

Coming to the insect world, we find a formidable gathering ready to oppress, rob, waylay, or slaughter the bee. Paradoxically, one of the bee's worst enemies is the bee itself (mankind is in a similar position, only more so). Selfless, loyal to the death in respect of its own colony, if opportunity offers, the bee will attack and massacre other colonies for the base purpose of loot—not for itself, of course, but for the communal stores of its own city.

But I have already dealt in some detail with robber bees, so will pass on to that pest of bee and beekeeper, that pirate and highwayman, that thief and brigand, that yellow curse—the wasp. By inclination the wasp is not a fighter, though when it does fight it is more than a match for any bee. It relies more on agility, combined, when necessary, with brute strength. At dodging and swerving it is in a class by itself—in its dealings with bees, that is. In its dealings with man it is rather at a loss. Picnickers may admire (or the reverse) its agility when they try to squash it with a spoon, and they may crow when they pin this artful dodger with a knife brought slowly down, but whether they get it or not, they do not witness the wasp's true form. To witness this they must see it dodging a line of oncoming bees and getting into a hive, and out again fully laden. With man the wasp gets an inferiority complex. It may not convey that impression when it is stealing the jam even as we lift it to our lips, but I am sure it has. Those sudden movements upset it; all in all it deals very adroitly with descending spoons and knives, but it cannot anticipate every movement of its adversary as it does when dealing with bees.

With men wasps are not aggressive—at least, I rarely found them so, and I used to chivy them about a lot. Treatment that would have brought blind fury to a bee left the wasp still without any desire to get its own back. But there was one summer when the wasps were particularly bad. To give some idea of their phenomenal numbers I may state that they ate a considerable portion of my study—or rather chewed it off and took it away for their nests. I had a large outside study in the orchard, made of Columbian cedar, which is a strong, light wood that requires no paint or preservatives. Even the roof was tiled with wooden tiles of cedar. In previous seasons the wasps had not touched it, but this year there was not a square inch left ungnawed. The whole outside appearance of the study was altered; the smooth finish of the wood had given place to a mat, uneven surface like the surface of a slab of cheese gnawed by mice.

I had been all over the countryside with a tin of cyanide destroying scores of nests. It seemed to make no difference; the wasps came in ever-increasing numbers to the hives. Every morning the bees reinforced their guards. Phalanxes of them turned out, cold and miserable and apprehensive, soon after it was light. But the wasps were always there first. The bee is not naturally an early riser, and by the time the guards had got warmed up wasps were tottering down the alighting board heavy and almost helpless with the honey they had filched. As the day wore on things got worse. The foragers returning with full cargoes never knew if they would be able to make the short journey up the alighting board to the entrance without being cut in two for the sake of their laden abdomens. The ground in front of the hives became littered with the top halves of severed bees. And I could do nothing about it. I laid the

usual bottle traps baited with beer and other stuff, which caught a
few—in fact, a lot—but it made no difference. A feeling of baffled
fury possessed me; for I loved my bees and liked to see them working, happy and uninterrupted. Going to the honey shed one morning I found that the wasps had discovered that also. They had got
at some honey that was stored inside, and great was their delight.
The place was stiff with them. I got busy with a swatter, and no
doubt wielded it in a manner that gave some expression to my pent-
up feelings. I mowed them down and they fell like corn before the
sickle. Then one stung me on the neck. I went on with my work,
but that sting was a signal. A second later, all those that were left—
and there were still many—came for me. It was a striking transformation; one moment their only ambition had been to get away
from me, the next to get at me. I was enveloped in a cloud of them
when I turned and slammed the door and ran. I was soon back,
this time with a veil and a tin of Flit. I found them still in the same
mind, but I was now able to cope with them. That is the only
occasion I have ever known wasps really to lose their tempers and
to act in concert, and I have waged war on them for many years.
This incident seems to me another illustration of that mysterious
power possessed by social insects for communicating news or states
of mind to their fellows. All these wasps may well have come from
the same nest. One of them, after perpetual chivying, changed
its attitude from panic to anger and this change was, almost immediately, communicated to the rest.

Luckily for bees, wasps are omnivorous and live largely on insects, and insects in a masticated form are the diet of their young,
so that while some are plundering the hives others are out hunting.
Were entire colonies of wasps to make concerted attacks on a hive
—as bees do—the hive could not withstand them.

It is a rather dreadful world into which the bee launches itself when it leaves the hive on its proper occasions. The insect world is bad enough for us, with creatures to bite, sting, and torment us, poison our food, give us diseases, eat our crops, clothes and houses, and even live on our bodies, but we are spared the ordeal of having them strike us down and devour us. We can, for instance, instead of fleeing for our lives, stay and admire the iridescent colors and darting flight of that voracious rover of the air, the dragonfly or devil's-darning-needle. But to an insect it means death. Those it has marked cannot escape; it is too swift. Even the wasp's dodging will not help, nor will its strength: it will be torn to pieces like a lamb by a hyena. In this it is but meeting with its deserts, but the unoffending bee gets the same treatment. How many bees the dragonfly takes it is impossible to say. The bees go forth, and in the evening so many fail to return. Which enemy has accounted for them can never be known.

Half hidden at the edge of her frail and diabolical creation of woven silk waits the spider. The spider's web, strong as it is—and for a thing so tenuous it is very strong—could no more hold a normal bee than the netting we put over raspberries could hold a mastiff. Yet the spider snares and kills a fair number. Once, among the brambles, I saw a dead bee in a spider's web, and curiosity prompted me to make a search. The brambles were full of webs and about one in five held a bee. There was a large area of brambles, so the mortality amongst the bees must have been heavy. True, they were mostly old bees, but to their communities they were cargo vessels carrying a full load and capable—but for the spider—of carrying many more. It is important—if one is an insect—to get out of a spider's web quickly. If one fails at the first attempt and takes a rest, the spider runs out and bites and then runs back again. The bite, as a bite, is nothing much; but poison that will bring partial paralysis has been injected. The spider, if she wished, could bite in certain spots which would bring death speedily. But she does not wish; web-weaving spiders either bind their prey or benumb it. They do not want it dead; they wish to suck the living blood.

Spiders' webs are found everywhere, but they are more numerous in brambles than anywhere else. At certain seasons the bees work the brambles, and very exacting work it is. To get to the majority of the flowers they must penetrate into thickets and force their way through rough leaves and prickly branches. This tears their coats and frays their wings and tires them. The first web a bee goes into merely amounts to a mishap for the spider; she must make her web anew. The bee gets out, goes home with her load, and comes back again. At the end of a day of honey-gathering any bee is tired, and if she has been working brambles she is exhausted. It is during those last journeys of the day that she falls a victim. Perhaps she fights a long, hard fight in a web, and, getting free at last, blunders straightway into another. This is the end. There is no more strength left in her. Her course is run, and the spider sucks her blood. There is another spider, a crab spider, which does not weave a web but hides up in flowers. When a bee arrives and becomes occupied in getting nectar, this spider springs on it, bites it on the neck, and kills it almost immediately.

Somebody, sometime, must have started the rumor that the Death's-Head moth is one of the greatest of all enemies of bees. At the beginning of my efforts at beekeeping, when I read every book or pamphlet, ancient or modern, about bees that I could get hold of, I remember being most impressed by a writer who described dramatically the paralysis of fear that came upon the bees when this dreaded visitor approached. What it did, or intended to do, when inside was not mentioned. If hives really did attract this rare and valuable moth in any numbers, beekeeping would be a more paying proposition than it is today.

Ants come into the list, though in talking about ants we are on rather difficult ground. One cannot lump them together; for the various and innumerable species differ so much the one from the other in character and mode of life that at the ends of the scale we have creatures as divergent as a rabbit from a man-eating tiger. The important thing, for the bees, is to keep away the single ants, the wandering scouts. If one of these gets inside, and out again, others will come, and soon there will be a regular column marching to and from the hive, with which the bees are quite unable to cope. They cannot sting these minute creatures, and it would make little difference if they could; they can only worry some of them and try to pull them away, and while they are doing this scores of others are marching past. It is not as a rule the bee's fault. She keeps the wandering scout away from her front door, but the ants are small, and the modern hive is full of little crevices. It is through these that the columns go. Provided no grass is allowed to touch the hive the scouts can be kept away by cups of water or paraffin under the hive legs, or bands of sticky paper wrapped round them, but once a real procession has started, the ants usually manage to fill in the one or bridge the other with their dead bodies.

Ants constitute a serious problem only when their nests are close and numerous. Ordinarily, the nests can be found and destroyed. The bulk of their looting takes place before they are discovered. The beekeeper studies the front of the hive; that little brown column working so industriously through a chink at the back escapes notice.

On summer evenings a small, grayish moth may sometimes be seen hovering about the hives. It dances and flutters first in front of one hive, and then another. Whether or not it is studying the number and disposition of the guards I do not know, but suddenly it drops on to the alighting board of one of the hives and with a couple of twists and turns dodges through the aperture. The guards look a bit bewildered and run about questingly, but soon other suspicious characters engage their attention and the incident passes from their minds. The wax moth is now inside and is mingling boldly with a crowd of bees who, for the present, pay not the slightest attention. I admire the way bees concentrate on their work, but I wish they would occasionally pay a little more attention to the shady characters that get in; they would save themselves a lot of trouble later.

The wax moth, brazen as it is, does not stay too long among these preoccupied creatures; it wanders away to the quiet, secluded parts of the hive, and in the bits of wax on the frame tops, and in the corners and on the floorboards, lays eggs. After that I suppose it dodges its way out again, though I have never seen it. Soon the eggs hatch out and a little army of minute grubs burrows tunnels all along the combs in the midriff. They line these tunnels with silk so that the bees find it difficult to get at them, and as they tunnel they eat wax, pollen, honey, and, at times, the babies in the nurseries. Even so, it is difficult to understand how their race endured before man started beekeeping. They are troublesome creatures to deal with, but the bees *do* deal with them. In a normal hive not one grub lives to perpetuate its species.

The movable-comb hive, however, allows the beekeeper to store combs from season to season. In time he usually accumulates a large number, half of them black, moldy, and fit only for the fire. Yet, as a rule, he will no more part with even the worst of them than a miser will part with his coins. This is a harmless idiosyncrasy provided he stores them securely: if he does not, if he leaves the smallest aperture, the wax moth will get in and reduce the lot to riddled, powdery wreckage. This may be the best thing that could happen to many of the combs, but it means a big increase later in the wax moth population and a lot more work for the bees.

Finally, mention must be made of parasites, but I will not dwell on this unpleasant subject. Bees are the cleanest of all creatures, but they are by no means immune. There is a fly in Africa something like a horsefly, and this attacks human beings. It only gives a slight "bite," which among so many biting insects is not particularly noticed. It does not really bite at all, but lays an egg under the skin, which hatches and develops into a maggot of about the size of a peanut. There is nothing to be done about it until the maggot is ready to emerge, when what is now a large abscess bursts. Having had one in my arm I can vouch for the intense and increasing agony after the egg hatches out and the maggot moves about inside. There are two known flies that attack bees in exactly the same way, though the maggot is correspondingly larger and causes the death of the bee, probably from pain and wasting, before it emerges. The bee also has a louse—or what is called a bee louse (its other name is *Braula coeca*). It lays its eggs in the combs and only fastens itself (very firmly) onto the bee when full grown. It is about the size of a pin's head, which would correspond to something the size of a crab with us.

In examining this list—which is far from complete—two facts, I think, stand out: (1) that the bee has more enemies, and from a wider class, than any other creature; and (2) that the majority of these enemies are such only because the bee stores honey. In restricting themselves solely to a diet of the essence of certain flowers and their pollen, which is available only for a few months in the year and not always then, bees, it seems to me, have set themselves an unnecessary handicap. Few other social insects have such a limitation; there is, for instance, very little that ants will not eat, and if we include the termites, practically nothing at all except stone and metal. Personally, I am aghast at the hazardous existence bees lead, owing to the cupidity this food of theirs excites in others, and sometimes I wish that they did not store up honey at all. Speaking from a selfish point of view, we could do without it now that sugar is available; but we could not do without fruit, and except the bees work the flowers, the fruit tree blossoms but in vain.

CHAPTER TEN

The Last Load

As I have said already, my first hive of bees fascinated me so much that I could hardly tear myself from them. I felt I wanted to get to *know* them. Their indifference to me (so strikingly at variance with the attitude of their African relatives) only made me the keener. They must have known me by now. They could not have helped it. I was forever standing there. But they showed no consciousness of it. I can imagine a young bee, out for the first time, spotting me and saying to another, "I say! Look! There's a great animal standing by the hive!" And the other, "Oh, that! Don't take any notice of it. It's always been there." But to outward ap-

pearance they were unaware of my existence. So, being unable to
get in touch with them, I lived with them in imagination; or,
rather, I selected one bee and lived with her.

This bee of mine I saw emerge with painful struggles from her
cell: soft, gray, velvety, and stingless. I saw her wandering upon
the comb, knocked over without apology by every passing porter.
Later I saw her trying to get a job of work. But they would have
none of her. She was too young; her "bones" * had not set. She
need not worry. There will be work in plenty soon enough: there
will rarely be anything else. For two days she lives in idleness; in
luxury, too, for the honey vats are at her disposal, she can sip when
she likes and she is not above soliciting passing workers to feed her.
Then one morning I see her proudly busy on her first job. It is no
exalted job; she is cleaning out brood cells and removing any
excreta left by the last occupant. It may be lowly work, but it must
be done properly. After cleaning, the cell must be polished—
polished till it shines again like a house-proud woman's drawing-
room furniture—for no queen will lay an egg in any cell not thus
polished. A gang of scavengers are working in collaboration below.
Each speck of dirt she brushes out is seized and carried from the
hive. It is not dumped on the threshold; bees are not like that. It is
carried to the next field, or even farther. This cleaning and polish-
ing is hard on a young child, and my bee is not tried beyond her
strength. Every now and then she is set to work on incubating
eggs—the only "soft job" in the hive, and reserved for the very
young, as a rest.

She is launched now. She is a working member of the hive. She

---

* The "bones" of the bee are outside. Her covering hardens in a few
days and becomes like a suit or armor called "chitin." Unlike ourselves she has no healing apparatus. Any wound remains: there is no
repair.

will get changes and adventures in plenty. There is only one thing she will never know, and that is rest. * Others are being born. Soft, downy bees in numbers are being put on their first work. In three days my bee receives orders. She is to report on the morrow at the nursing ward for instructions. She is thrilled, of course. Bees *are* thrilled with their work. You have only to watch them to realize it. And those wise overseers in the hive are careful never to keep them on one task too long.

So my bee reports to the nursing ward and soon is busy feeding the old grubs. Mark the word "old." Were she and her kind to feed the very young grubs things would go wrong. She has not the experience yet to feed the real babies and day-olds, nor, and more important, the means. Young grubs have to be fed for the first three days on a kind of mother's milk, though sister's milk is a more correct description. This milk comes from glands in the workers' heads which at certain times secrete a fluid rich in proteins and vitamins. In spite of all the care taken things often go wrong; young worker grubs may be fed for too long on this milk and develop later into "laying workers," laying eggs that develop into drones. No very serious matter normally, except for the laying workers themselves, for they are soon killed off, but serious at certain times such as sudden queenlessness when the presence of eggs may possibly give a colony the impression that they have a queen when, in fact, they have not, and may delay their preparations for raising a new queen until it is too late.

Meanwhile, my bee, in-between feeding the old grubs on a mixture of watery honey and pollen, has been continually creeping off to the pollen cells, those tightly-packed tubs of flour, and guzzling

---

* Except when she is a forager and weather conditions make it impossible to go out to the fields.

at them. Like a pregnant mother she feels the urge for unusual foods. It is her glands that are demanding this pollen which is converted into a rich milk for young grubs. These milk glands develop at about the sixth day and fade by the twelfth day, at which time other glands begin to store wax. This does not mean that young grubs can only be fed with appropriate food by nurses of this age group as has sometimes been supposed. Old bees can do it too. A swarm, for instance, consists mostly of old bees, and even the young ones are old by the time the new combs have been built and the eggs laid in them hatched; yet these old bees can secrete the necessary milk and feed it to the young grubs. Everything in a hive has to be elastic; there can be no hard and fast rules, but with an established colony in spring and summer the duties of the workers are allocated in more or less a fixed sequence. *

And so my bee friend when she is about seven days old starts to feed the very young grubs, and about now she begins to go to school also. Her work in the hive is important, but she must prepare for the time when her duties are outside the hive. She must learn to fly. Classes are held twice a day in fine weather and last about half an hour. By relays the scholars jump and hover just before the entrance, then sit for a while in the sun and watch the traffic of foragers above and wish no doubt that they, too, could fly so swiftly and surely and strongly.

Soon she can fly; not with the swift and certain flight of the foragers, but sufficiently well. Once flying is mastered she must

---

* The various duties of bees and the age at which they engage on them have been investigated by research workers with the aid of marked bees. The ages vary according to the exigencies of the hive. Investigation is not yet complete. There are many duties not mentioned here, and probably a lot of drafting and redrafting of gangs takes place according to necessity.

acquire another accomplishment: she must use her brain and
memory and orientate the hive, for a bee gets back home not by a
sense of direction like a cat or a pigeon, but in the same way that
a hunter gets back to his camp—by noting landmarks on the way
out. Take away her landmarks (trees, barns, hedges—anything
serves) and the bee is lost. Or take her a short distance outside the
country she has learned and she will never get home. At school she
learns the landmarks near the hive, and every day she learns others
a little farther off. But school, important as it is, is only an inter-
lude; work is the chief thing in the bee regime.

Nursing is exhausting work (it has been estimated that the rearing of a single larva entails between two and three thousand visits
by attendants) and so important that the bees run no risk of having
stale nurses. My friend is taken off in about four days and enters a
more humble sphere: she becomes a porter and maid-of-all-work.
She helps the foragers with their loads when they arrive, digests
honey, rams down the pollen in the pollen cells, removes litter and
corpses from the hive, and joins the ventilation and honey-brewing
gangs. This work is merely given her as a sort of break. She has been
on one important task and is shortly to go on another. It is the bees'
idea of a rest. In three days she is ready to engage in that most intricate and delicate of all work—comb building. Obviously some
young bees may never be called upon for this, and others may have
only an odd repairing job or so. But a swarm has a whole city to
build. I cannot follow her here; her duties have become too complicated. The architects will have mapped out the position and studied
stresses and weights and all the rest of it. A great cluster forms, in
which—somewhere—is my girl friend.

The hexagonal comb of the bee has aroused interest even among
mathematicians. Wasps, of course, build combs, quite neat combs
in layers resting on pillars. But there is a vast difference between a

one-sided horizontal slab and a suspended *double* comb. No other insect can build a double comb. It is a marvel of intricate precision. Examine a piece of bees' comb for yourself. See how the apex of the base of one cell forms part of the upper facet of the cell on the other side. Consider the geometrical exactitude required and note the carefully thought out saving of space. The newly finished product is white and exceedingly fragile. A piece a foot square weighs little more than a feather. A touch will crush the cell sides. Yet that small piece will bear the weight of eight pounds of honey as it hangs suspended.

For six days my friend labors, one of a long black mass being gradually lowered on a curtain of glistening white. Bees, as we know, work like slaves, but they never have to complain of monotony. The self-same routine of human factory and office workers is not for them. Even the foragers get the thrills of search and discovery. All the same, the job they give my bee now is at startling variance with what she has done before. She becomes a guard! I feel a little uneasy about this. A guard's post is no sinecure. Orders insist that any marauder must be attacked on sight. The guard that spots an intruder must go for it without waiting to sound the alarm. (Those bees that occasionally shoot at you and sting you when you make an abrupt movement near a hive are guards.) And unless the hive is attacked by numbers fellow guards will not concern themselves with another's fight. My friend, engaged in a life and death struggle with a robber bee, would in all probability be left to fight it out alone. Hives differ. Some guards combine more than others. In any case a robber bee seems to escape more easily from three guards than it does from one. Three guards get in each other's way.

The dreaded enemy is the wasp. Larger, more active, more powerful than the bee, he takes a lot of stopping. In single combat the bee has no earthly chance. The sight of his approaching form causes obvious uneasiness among the guards. They show an inclination to edge away, each hoping another will be the one to sacrifice her life in the first fierce struggle. But it has to be done. Once a wasp has forced his way inside and got out with a load of honey he will come back. And when he comes back he will treat the guards with supreme contempt, shouldering them aside and marching in as if they did not exist. And he will bring friends with him. There is a swaggering bravado about the wasp to which one must accord a grudging admiration. There is a D'Artagnan quality in the light-hearted way he takes on heavy odds. Whether he can get in or not, he haunts the hive. And if he cannot get honey from the combs, he gets it from the bees. His unpleasant method is to pounce on tired returning foragers, bite them in half on their own doorstep, and dash off with the loaded abdomen before a guard can get near him. He is a buccaneer, a cruel and sinister pirate and well termed the "yellow peril."

In observation hives I have studied many a wasp after he has fought or outwitted the guards and got inside. In such circumstances a robber bee is unnoticed, but the wasp remains the hated enemy. Strength and quickness, particularly quickness, are what he relies on. At first sight it would seem that a single wasp venturing into such a place was merely committing suicide. But where he gets away with it (when he does) is through his unexpectedness. To get a true picture you must imagine a tiger appearing in a warehouse full of clerks, porters, carpenters and the rest, *all* utterly engrossed in their work—though you may have to strain your imagination a little for this! Before those in its vicinity have realized it is there, the tiger has snatched up some booty and bounded off to a different part of the building. And before the place as a whole is aware a tiger is there at all, the tiger has gone, with as much as it can carry, and is making for home.

The feverish rapidity with which a wasp pumps honey from a cell, dodges an onrushing crowd, then pumps honey from another cell is more exciting to watch than any football match. You witness more skill and judgment. And to add to the interest you realize that the wasp is playing with death, and knows it. The slightest miscalculation, the smallest delay, and his number is up. Not infrequently his greed prompts him to pump away at a honey cell just a little too long. Withdrawing, he finds himself surrounded by a larger force than even he can throw off. A small black ball falls to the floor of the hive—a knot of bees in the center of which is the madly fighting wasp. Sting or bite as he likes now, he has run his course.

It is a different matter when his exit is barred. Sometimes, when a wasp has entered, I have clapped a piece of perforated zinc over the hive entrance. What will the robber do now— trapped, with forty thousand gradually growing aware of his presence? When he makes his final dash and finds his retreat cut off, terror enters his fierce soul. He runs about feverishly seeking another exit. His strength suffices as yet to brush all opposing bees aside. They bide their time, and as he weakens fall on him. The bold bad robber dies, and I remove the zinc so that they can take his body out, and the workers, thronging and fuming at the barrier, can get on with their jobs.

No wonder I am glad when my friend's period of military service expires and she enters upon her final calling and at long last becomes a forager. This is the goal of every bee; the crowning sign of adolescence. It is a bright morning when she emerges to take her first working flight. I watch her as she stands there giving herself a final clean and shake before launching off. She is good-looking;

larger and handsomer than the bulk of the foragers with their frayed wings and dark, work-worn bodies. Her delicate coat of down is new and shining; her orange bands are brilliant; her wings, perfectly shaped, glisten in the sun. Certainly a good-looker is my girl friend.

She is off. The patent self-locking teeth snap together, joining the double wings into one rigid whole, and, soaring aloft to get her bearings, she disappears. She knows where she is going. On her school flights she has seen the campanulas growing in the sunny border before my study window. She descends slowly: it is the first time she has ever alighted, except on the well-known landing place in front of the hive. The flower bends slightly as she falls and clings to the blue, turned-down edge. It is heaven inside; an intoxicating perfume. Rudely she brushes past the floured barriers and licks a bit of sticky sweet stuff at the base. It is not honey, just a faint sweet residue. Dust is falling all about her; delightful, scented dust. She wants to roll in it; bathe in it.

She leaves the flower at last. She can hardly fly. Every part of her is a mass of dust. She hovers over the bed, combing and cleaning herself and cramming the combings into two baskets on her hind legs. Gradually she gets clean, and being clean longs for another bath in some more warm, scented dust. She descends once more. Another flower bows and receives her. Flower after flower she visits, and then she tries a poppy which a lot of other bees seem to be finding irresistible. The scent revolts her. It does not "mix" at all with the scent that now pervades her body. She returns to the campanulas.

Rising from a flower she is seized with alarm. Her hind legs are leaden weights. She can hardly fly. Will she ever get home? Slowly she climbs and makes for the hive. It comes in view at last. But can she make it? Easy enough in the old school days, but very different when she is weighted down by the mightiest load of pollen bee ever bore.

She makes her landing and proudly paddles her yellow loads towards the aperture. You or I would have seen a bee arriving with a negligible load of pollen on each leg. My friend (my human friend) would have said curtly, "Young bee. First pollen trip." And he would have said it with a kind of contempt. But the other bees do not sneer like that. They understand the pride which fills my bee. She is bringing home her *first load!*

They know that the life of the hive depends on the enthusiasm of the foragers. Whether there are overseers in the hive or not, there are none in the fields. So they always make a great fuss of these pathetic first attempts. They make no fuss of her afterwards when she comes back regularly with bulging thighs. Perhaps enthusiasm is important, psychologically, at the start. It is so with our youngsters: why not with theirs? I like to think of this first reception. It throws new light on the bee, supposedly so callous and machinelike. My bee is longing for praise, and she gets it. Bees come running out and crowd round her. In an admiring circle they escort her to the combs, and my bee's cup of joy is full. She watches her precious but microscopic load rammed neatly into the pollen cells (making no appreciable difference to their capacity) and is off again, singing, to the sunny corner by my study window. In another day or two she is just another of that vast army of pollen bearers coming home with their loads every few hours.

These first days of foraging are very pleasant to the bee. She is equipped for it, made for it, and for the first time in her life she makes the acquaintance of the bees' mutually dependent partners—the flowers. The two are wedded. It was a marriage consummated in a dim and distant past, but they have been faithful ever since. Neither could survive without the other. It is no

Platonic union: the virgin stigma of the blossoms knows nothing of the distant anthers. The bee is *her* husband. He comes to her lusty and pollen coated, and is received with open arms. It is a happy, fragrant marriage.

The pollen gatherer gets all the applause of human onlookers. Her load is visual, and she seems to carry by far the biggest cargoes into the hive. As a matter of fact she does nothing of the sort. However large her burden, it is nothing compared to the weight of a honey forager. That is why my bee was set to bring in pollen first. Her wings are not strong enough nor her flying experience great enough to navigate such a heavy load as honey. But daily she gains in strength and knowledge. She can bring home her pollen loads in all conditions, against, or, what is harder, with the wind. She is promoted to her final task. She gathers honey.

She has much to learn. She knows the country for a radius of two miles from the hive, but she does not know the flowers—not in respect of nectar yielding. Flowers give pollen in abundance, but they are more chary about giving nectar. The yielding of flowers is like the "rise" of trout. It comes suddenly at certain times in certain places, and ceases as abruptly. Any fool of a bee can get honey when the clover is yielding everywhere, but it takes a wise bee with *knowledge* to get a load when only a few flowers in certain places give a reluctant supply. And there is the right of priority to consider. Going through a field of clover in yield you will note that although it is full of bees there is never any crowding. The law is that if any small patch has a certain number of bees at work a newcomer must go elsewhere. This has been proved by the experiments of an investigator, whose name I am afraid I forget. So you see my bee is going to have to work hard and fly long distances to get her regular loads of honey. Nevertheless, her ways are pleasant ways. It may be hard, often, to get a load, but the thrill of bringing one home never lessens. There is zest in the life; the zest of discovery and of competition and of unexpected "strikes." And there is the joy of certain days when nature is smiling and honey to be had for the getting. It is all hard work, but it is all joy. And she is young. An hour's rest in the hive after each journey brings her out again as fresh as ever.

Truth to tell, she is not so good-looking as she was. Pushing through grass and leaves and the twigs of brambles and wild thyme has played havoc with her smart fur coat. There is little of it left. She looks smaller and shinier, and her wings are frayed at the edges. I regret this particularly. I did so admire the perfect shape of those transparent wings. It does not worry *her*. She can fly with the best, and that is all she cares about. In any case the bee flies in defiance of the laws of gravity. According to experts she ought not to be able to fly at all: her wing area is too small. But she does, and furthermore is one of the swiftest flyers we know. When the wings become frayed she has only to move them a little faster to fly as quickly as she did before. Perhaps my bee, when heavily laden, finds the return journey more laborious than it used to be, that is all.

Day follows day, and regularly my friend appears with her load of honey, toiling up the alighting board. But the pace is fast. The wings, merely frayed wings once, are torn wings now. Her body is dark and looks "bare." My beautiful young bee is beautiful no longer. Nor is she young; for age in workers is reckoned by wear and tear and not by days. Time was, not many days ago, when she was reckoned one of the lights of the hive; she brought in heavier loads more often than the others. It is not so now. Her flower lore has increased, but the journeys take so long. Her landings are getting clumsy, too. She lacks control; her wings do not bite the air as they did.

A few days later they begin to work the brambles in earnest. Brambles are hard on wings, and, moreover, are full of spiders' webs. My friend today gets caught in several. She breaks free, but the last one holds her a long time. And in the eyes of the spider beneath a leaf was a look of speculation. No spider had looked at her quite like that before. It was after mid-day when she returned. What joy to be back—home in the familiar roar and heat and thronging crowds, all so busy, pushing her aside. A young bee, a very young bee, resplendent in shining coat and brilliant bands, receives her and leads her to a cell and helps her discharge her load . . . even as she herself did when an eager youngster. She rests a little. She would have liked to have rested longer, but the day is far gone and she has brought in only one load. She appears at the entrance again in a quarter of an hour: a little blackish thing with stumps of wings. My poor little bee! How can you fly at all!

She has an idea that a patch of charlock, dry as a bone that morning, may possibly have begun to yield after this hot sun. She is right. And, moreover, she is almost the first on the scene. She is back at the hive quite soon with a very respectable cargo, and with her self-respect restored. She falls short, as is so usual now, but after a panting rest on a stone she makes the entrance. There is still time for one more load. She will not have done too badly. She hurries off for the last load.

The patch of charlock is occupied. It hums with bees. A field of mustard is the most likely alternative. Unfortunately it is a mile away. She is panting when she gets there, unladen though she is. Again she is right. Her intuition in the matter of flowers is almost uncanny. She works quickly. A chill in the air warns her that night is approaching. She rises from the last flower and makes for home. She is vaguely troubled with the thought of the long hill which must be surmounted. After that it will be all descent to the hive. The hill seems far away, but at last it looms before her. She mounts, or tries to, but gains no altitude. With all her energy and will power she vibrates her stumps of wings. The earth comes toward her. She alights and clings to a blade of grass.

A cold drizzle falls that night, but she is still clinging there next morning. With remarkable tenacity she holds on for two more rainy days. Then, on a warm, sunny morning, just at the time the youngsters at the hive are coming out for school, her grip relaxes and she falls into the wet grass below.